Deep Learning Approach for Natural Language Processing, Speech, and Computer Vision

Deep Learning Approach for Natural Language Processing, Speech, and Computer Vision provides an overview of general deep learning methodology and its applications of natural language processing (NLP), speech, and computer vision tasks. It simplifies and presents the concepts of deep learning in a comprehensive manner, with suitable, full-fledged examples of deep learning models, with an aim to bridge the gap between the theoretical and the applications using case studies with code, experiments, and supporting analysis.

Features:

- Covers latest developments in deep learning techniques as applied to audio analysis, computer vision, and natural language processing.
- Introduces contemporary applications of deep learning techniques as applied to audio, textual, and visual processing.
- Discovers deep learning frameworks and libraries for NLP, speech, and computer vision in Python.
- Gives insights into using the tools and libraries in Python for real-world applications.
- Provides easily accessible tutorials and real-world case studies with code to provide hands-on experience.

This book is aimed at researchers and graduate students in computer engineering, image, speech, and text processing.

Deep Learning Approach for Natural Language Processing, Speech, and Computer Vision

Techniques and Use Cases

L. Ashok Kumar and D. Karthika Renuka

CRC Press
Taylor & Francis Group
Boca Raton London

CRC Press is an imprint of the
Taylor & Francis Group, an **informa** business

First edition published 2023
by CRC Press
6000 Broken Sound Parkway NW, Suite 300, Boca Raton, FL 33487–2742

and by CRC Press
4 Park Square, Milton Park, Abingdon, Oxon, OX14 4RN

CRC Press is an imprint of Taylor & Francis Group, LLC

© 2023 L. Ashok Kumar and D. Karthika Renuka

ISBN: 978-1-032-39165-6 (hbk)
ISBN: 978-1-032-39166-3 (pbk)
ISBN: 978-1-003-34868-9 (ebk)

DOI: 10.1201/9781003348689

Typeset in Times
by Apex CoVantage, LLC

Dedication

To my wife Ms. Y. Uma Maheswari and daughter
A. K. Sangamithra, for their constant support and love.

—Dr. L. Ashok Kumar

To my parents Mr. N. Dhanaraj and Ms. D. Anuradha, who laid the foundation for
all my success, to my husband Mr. R. Sathish Kumar and my daughter P.S. Preethi
for their unconditional love and support for competition of this book, and to my
friend, brother and co-author of this book Dr. L. Ashok Kumar, who has been highly
credible and my greatest source of motivation and as a pillar of inspiration.

—Dr. D. Karthika Renuka

Contents

About the Authors

L. Ashok Kumar was Postdoctoral Research Fellow from San Diego State University, California. He was selected among seven scientists in India for the BHAVAN Fellowship from the Indo-US Science and Technology Forum, and, also, he received SYST Fellowship from DST, Government of India. He has 3 years of industrial experience and 22 years of academic and research experience. He has published 173 technical papers in international and national journals and presented 167 papers in national and international conferences. He has completed 26 Government-of-India-funded projects worth about 15 crores, and currently 9 projects worth about 12 crores are in progress. He has developed 27 products, and out of that 23 products have been technology-transferred to industries and for government-funding agencies. He has created six Centres of Excellence at PSG College of Technology in collaboration with government agencies and industries, namely, Centre for Audio Visual Speech Recognition, Centre for Alternate Cooling Technologies, Centre for Industrial Cyber Physical Systems Research Centre for Excellence in LV Switchgear, Centre for Renewable Energy Systems, Centre for Excellence in Solar PV Systems, and Centre for Excellence in Solar Thermal Systems. His PhD work on wearable electronics earned him a national award from ISTE, and he has received 26 awards in national and in international levels. He has guided 92 graduate and postgraduate projects. He has produced 6 PhD scholars, and 12 candidates are doing PhD under his supervision. He has visited many countries for institute industry collaboration and as a keynote speaker. He has been an invited speaker in 345 programs. Also, he has organized 102 events, including conferences, workshops, and seminars. He completed his graduate program in Electrical and Electronics Engineering from University of Madras and his post-graduate from PSG College of Technology, Coimbatore, India, and Master's in Business Administration from IGNOU, New Delhi. After the completion of his graduate degree, he joined as Project Engineer for Serval Paper Boards Ltd. Coimbatore (now ITC Unit, Kova). Presently, he is working as Professor in the Department of EEE, PSG College of Technology. He is also Certified Chartered Engineer and BSI Certified ISO 50001 2008 Lead Auditor. He has authored 19 books in his areas of interest published by Springer, CRC Press, Elsevier, Nova Publishers, Cambridge University Press, Wiley, Lambert Publishing, and IGI Global. He has 11 patents, one design patent, and two copyrights to his credit and also contributed 18 chapters in various books. He is also Chairman of Indian Association of Energy Management Professionals and Executive Member in the Institution of Engineers, Coimbatore Executive Council Member in Institute of Smart Structure and Systems. Bangalore, and Associate Member in the Coimbatore District Small Industries Association (CODISSIA). He is also holding prestigious positions in various national and international forums, and he is Fellow Member in IET (UK), Fellow Member in IETE, Fellow Member in IE, and Senior Member in IEEE.

D. Karthika Renuka is Professor in the Department of Information Technology in PSG College of Technology. Her professional career of 18 years has been with

PSG College of Technology since 2004. She is Associate Dean (Students Welfare) and Convenor for the Students Welfare Committee in PSG College of Technology. She is a recipient of Indo-U.S. Fellowship for Women in STEMM (WISTEMM)— Women Overseas Fellowship program supported by the Department of Science and Technology (DST), Government of India, and implemented by the Indo-U.S. Science & Technology Forum (IUSSTF). She was Postdoctoral Research Fellow from Wright State University, Ohio, USA. Her area of specializations includes Data Mining, Evolutionary Algorithms, Soft Computing, Machine Learning and Deep Learning, Affective Computing, and Computer Vision. She has organized an international conference on Innovations in Computing Techniques on January 22–24, 2015 (ICICT2015), and national conference on "Information Processing and Remote Computing" on February 27 and 28, 2014 (NCIPRC 2014). Reviewer for Computers and Electrical Engineering for Elsevier and Wiley book chapter and Springer book chapters on "Knowledge Computing and its Applications," she is currently guiding eight research scholars for their PhD under Anna University, Tamil Nadu, Chennai. She has published several papers in reputed national and international journals and conferences.

Preface

In early days, applications were developed to establish an interaction between the humans and computer. In the current trend, humans have the most advanced methods of communications like text, speech, and images/video to interact with computers. Voice-based assistants, AI-based chatbots, and advanced driver assistance systems are examples of applications that are becoming more common in daily life. In particular, the profound success of deep learning in a wide variety of domains has served as a benchmark for the many downstream applications in artificial intelligence (AI). Application areas of AI include natural language processing (NLP), speech, and computer vision. The cutting-edge deep learning models have predominantly changed the perspectives of varied fields in AI, including speech, vision, and NLP. In this book, we made an attempt to explore the more recent developments of deep learning in the field of NLP, speech, and computer vision. With the knowledge in this book, the reader can understand the intuition behind the working of natural language applications, speech, and computer vision applications. NLP is a part of AI that makes computers to interpret the meaning of human language. NLP utilizes machine learning and deep learning algorithms to derive the context behind the raw text. Computer vision applications such as advanced driver assistance systems, augmented reality, virtual reality, and biometrics have advanced significantly. With the advances in deep learning and neural networks, the field of computer vision has made great strides in the last decade and now outperforms humans in tasks such as object detection and labeling. This book gives an easy understanding of the fundamental concepts of underlying deep learning algorithms to the students, researchers, and industrial researchers as well as anyone interested in deep learning and NLP. It serves as a source of motivation for those who want to create NLP, speech, and computer vision applications.

Acknowledgments

The authors are thankful to Shri L. Gopalakrishnan, Managing Trustee, PSG Institutions, and Dr. K. Prakasan, Principal, PSG College of Technology, Coimbatore, for their wholehearted cooperation and constant encouragement in this successful endeavor. The authors wish to acknowledge the Department of Science and Technology (DST) for sponsoring their project under DST-ICPS scheme which sowed the seeds among the authors in the area of Deep Learning Approach for Natural Language Processing, Speech, and Computer Vision. The authors thank the editorial team and the reviewers of CRC Press, Taylor & Francis Group for their relentless efforts in bringing out this book.

Dr. L. Ashok Kumar would like to take this opportunity to acknowledge the people who helped him in completing this book. He is thankful to his wife, Ms. Y. Uma Maheswari, and also grateful to his daughter, Ms. A. K. Sangamithra, for their constant support and care during writing.

Dr. D. Karthika Renuka would like to express gratitude to all her well-wishers and friends. She would also like to express her gratitude to her parents, Mr. N. Dhanaraj and Ms. D Anuradha, for their constant support. She gives heartfelt thanks to her husband, Mr. R. Sathish Kumar, and her dear daughter, Ms. P. S. Preethi, for their unconditional love which made her capable of achieving all her goals.

The authors are thankful to the almighty God for His immeasurable blessing upon their lives.

1 Introduction

LEARNING OUTCOMES

After reading this chapter, you will be able to:

- Understand the conventional machine learning task such as classification and regression.
- Understand the basic concepts behind speech, computer vision, and NLP.
- Identify the tools, techniques, and datasets used for speech, computer vision, and NLP applications.

1.1 INTRODUCTION

The fourth industrial revolution, according to the World Economic Forum, is about to begin. This will blend the physical and digital worlds in ways we couldn't imagine a few years ago. Advances in machine learning and AI will help usher in these existing changes. Machine learning is transformative which opens up new scenarios that were simply impossible a few years ago. Profound gaining addresses a significant change in perspective from customary programming improvement models. Instead of having to write explicit top down instructions for how software should behave, deep learning allows your software to generalize rules of operations. Deep learning models empower the engineers to configure, characterized by the information without the guidelines to compose. Deep learning models are conveyed at scale and creation applications—for example, car, gaming, medical services, and independent vehicles. Deep learning models employ artificial neural networks, which are computer architectures comprising multiple layers of interconnected components. By avoiding data transmission through these connected units, a neural network can learn how to approximate the computations required to transform inputs to outputs. Deep learning models require top-notch information to prepare a brain organization to carry out a particular errand. Contingent upon your expected applications, you might have to get thousands to millions of tests.

This chapter takes you on a journey of AI from where it got originated. It does not just involve the evolution of computer science, but it involves several fields say biology, statistics, and probability. Let us start its span from biological neurons; way back in 1871, Joseph von Gerlach proposed the reticulum theory, which asserted that "the nervous system is a single continuous network rather than a network of numerous separate cells." According to him, our human nervous system is a single system and not a network of discrete cells. Camillo Golgi was able to examine neural tissues in greater detail than ever before, thanks to a chemical reaction he discovered. He concluded that the human nervous system was composed of a single cell and reaffirmed

DOI: 10.1201/9781003348689-1

his support for the reticular theory. In 1888, Santiago Ramon y Cajal used Golgi's method to examine the nervous system and concluded that it is a collection of distinct cells rather than a single cell.

Way back in 1940, the first artificial neuron was proposed by McCulloch Pitts. Alan Turing proposed the theory of computation and also pioneered the idea of universal computer. Shannon came up with information theory which is extensively used in machine learning and signal processing. In 1950, Norbert Wiener is the originator of cybernetics. In 1951, Minsky developed the Neural Net machine called Stochastic Neural Automatic Reinforcement Calculator (SNARC). One of the most significant moments in the history of AI was the advancement of the primary electronic universally useful PC, known as the ENIAC (Electronic Numerical Integrator and Computer), which can be reprogrammed to tackle a wide range of numerical problems. Samuel created the first checkers-playing program for the IBM 701 in 1952. It has a memory of its previous game experience and applies the early experience in the current game. In 1955, Logic Theorist was the first AI program written by Allen Newell, Herbert A. Simon, and Cliff Shaw which is a proved theorem by Whitehead and Russell's *Principia Mathematica*. In the Dartmouth Conference in 1956, the term AI was coined with the aim to build thinking machines, and this event was the formal birth of AI. AI has the ability to change each and every individual in the world. Later, in 1957, Rosenblatt was the first to come up with an idea of two-layered artificial neural network called perceptron.

Arthur Samuel characterized AI in 1959 as the field of examination that permits PCs to learn without being expressly modified. The golden years of AI were between 1960 and 74 where the growth of expert systems happened, for example, when the idea of explicit, rule-based programs was born to address some problems like playing chess, understanding natural language, solving word problems in algebra, identifying infections, and recommending medications. But the expert system is not good enough as it is basically a rule-based model. In 1969, back propagation algorithm was derived by Bryson and Ho.

"Within twenty years, machines will be capable of accomplishing whatever work a man can do," predicted Nobel laureate H.A. Simon in 1965. In 1970, Marvin Minsky predicted that "Within three to eight years, we will have a machine with the general intelligence of an average human being." But actually, these two sayings were not true; 1974–80 is called the first winter of AI. It is because AI results were less addressed to real-world problems due to the lack of computational power, combinatorial explosion, and loss of government funding in AI.

The period 1980–2000 is said to be the spring of AI, when the first driverless car was invented. Minsky in 1984 said "winter is coming," and the period 1987–93 is the second AI winter due to lack of spectacular results and funding cuts. In 1997, IBM's Deep Blue beat world chess champion Kasparov in chess competition.

The current AI system is running inspired by the working of brain the same way airplanes are inspired by birds. The current AI system provides us a lot of excitement around how we can personalize applications to make smarter decision for people.

1.1.1 SUBSETS OF ARTIFICIAL INTELLIGENCE

Businesses can simplify operations by running AI models inside a process or application, using models to get predictions and insights by deploying AI-powered applications with the click of a mouse. AI models utilize less of the past data and predict more about the future. AI takes you to places you have only dreamed of.

We are filled with data around us, and big enterprises and organizations rely on digital technologies to support their work. In the past couple of decades, we have seen the growing evidence that AI-powered solutions have the potential to improve the business outcomes. With the advancement of AI-based techniques, massive amount of data and accelerated hardware performance greatly impacted almost all the sectors such as designing, health care, retail, production, space technology, and research. This section gives an outline of man-made reasoning, AI, and deep learning. The most general definition of AI is the use of computers or other devices to mimic human brain functions like problem solving and judgment. As illustrated in Figure 1.1, self-learning algorithms used in machine learning, a subset of AI, derive knowledge from data in order to forecast outcomes.

The two most common type of learning are supervised and unsupervised algorithms as depicted in Figure 1.2. Supervised learning utilizes named information to prepare calculations to group the information to foresee results. Labeled dataset means that the rows in the dataset are tagged or classified with some interesting information about the data. So, when to use supervised machine-learning techniques to attain your objective really depends on the use cases. For instance, classification model can be solved using supervised learning algorithm to classify the objects or ideas into predefined categories. A real-world example for the classification problem is customer retention. Based on the historical data on the activities of customers, we can assemble our own order model utilizing managed AI with our marked dataset, which will help us distinguish the clients that are about to churn and allow us

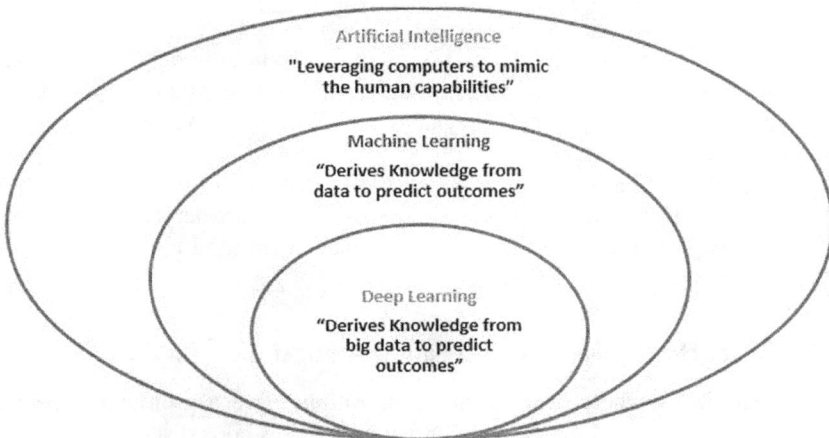

Artificial Intelligence
"Leveraging computers to mimic
the human capabilities"

Machine Learning
"Derives Knowledge from
data to predict outcomes"

Deep Learning
"Derives Knowledge from
big data to predict
outcomes"

FIGURE 1.1   Subsets of AI.

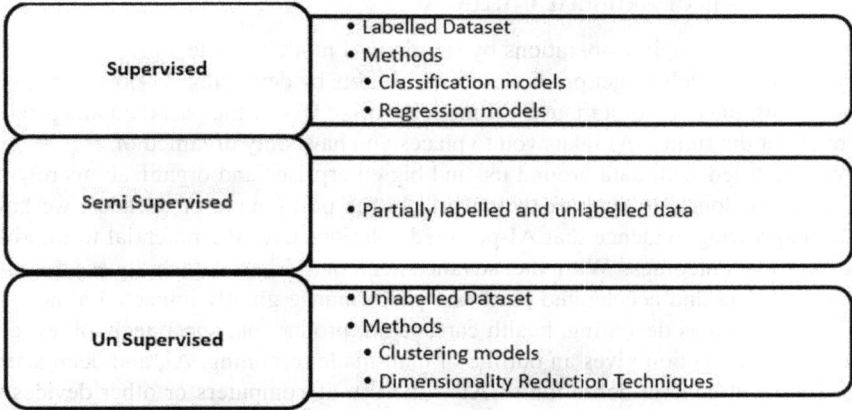

FIGURE 1.2   Types of Learning.

to make a move to hold them. Relapse is another kind of managed AI calculation, which is utilized to anticipate mathematical result values. For instance, airline companies rely more on machine learning algorithms, and they use regression technique to accurately predict things like the charge for a particular flight. Thus, the relapse examination utilizes different data sources, for example, days before takeoff, day of the week, source, and objectives to precisely anticipate the specific airfare that will amplify their income.

The second type of machine learning is unsupervised learning, which is used when the user wants to evaluate and cluster unlabeled data. Furthermore, this technique aids in the discovery of hidden patterns or groupings without the assistance of a human. Clustering model is a technique of unsupervised learning widely used for many use cases in the organization. For instance, if an organization wants to group their customers to do effective marketing, clustering mechanism helps to identify how certain customers are similar by taking into account a variety of information such as purchasing history, social media activity, and their geography. It groups similar customer into buckets, so that the organization can send them more relevant offers and provide them better service.

Deep learning, a subset of machine learning, predicts outcomes using massive amounts of data. The three forces that drive deep learning are depicted in Figure 1.3. This book discusses the various applications of deep learning algorithms in voice analytics, computer vision, and NLP.

1.1.2   THREE HORIZONS OF DEEP LEARNING APPLICATIONS

The three horizons where deep learning algorithms have a profound impact are shown in Figures 1.4 and 1.5. As shown in Figure 1.6, this book discusses the various applications of deep learning algorithms in computer vision, voice analytics, and NLP.

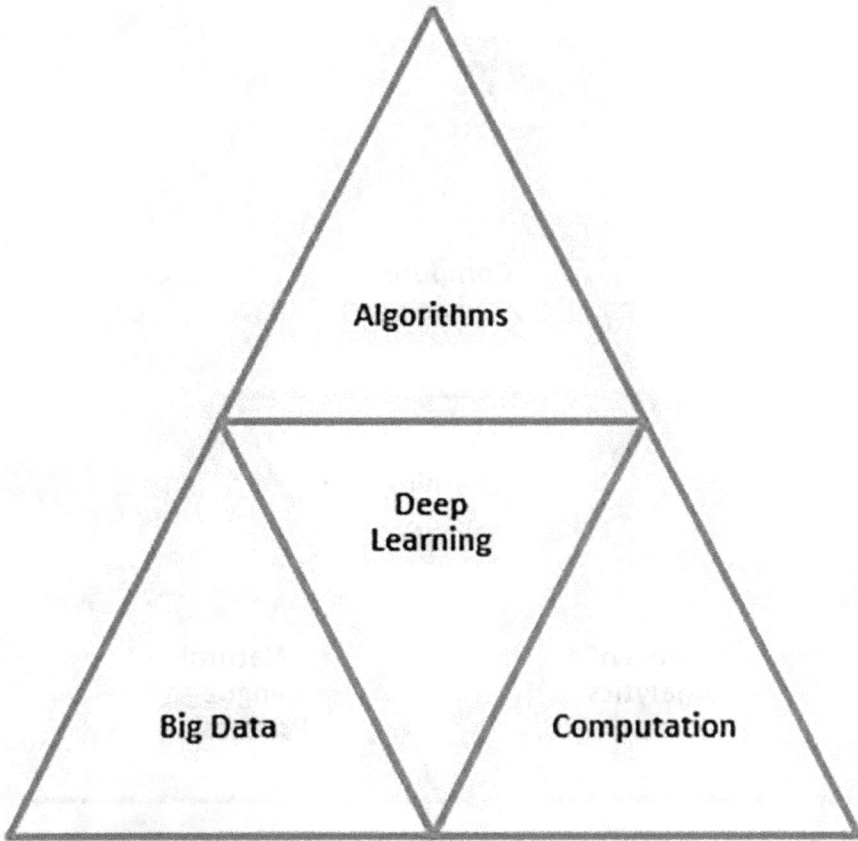

FIGURE 1.3   Deep Learning.

1.1.3 NATURAL LANGUAGE PROCESSING

Human beings effortlessly communicate and share information with each other. Human language is one of the most complex and diverse part. A huge amount of textual data is being generated by us in the form of tweets, WhatsApp messages, and social media comments. Most of the textual data is unstructured in nature, and, in order to get useful insights from them, techniques like text mining and NLP come into picture.

Text mining is defined as deriving useful insight from natural language text whereas NLP is an interdisciplinary field involving AI, Computational linguistics, Computer science which deals with human language as given in Figure 1.7. In recent times, NLP is gaining utility value in all sorts of AI applications. Figure 1.8 depicts the two variants of NLP.

- Natural Language Understanding (NLU): NLU is a process of converting unstructured messy human language into structured computer code.

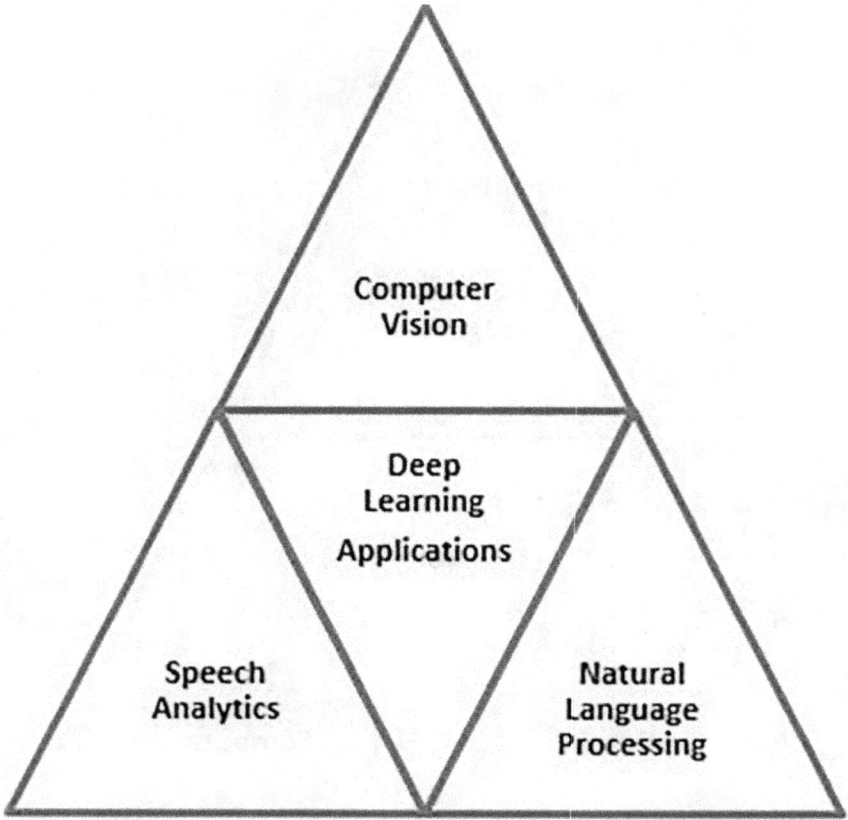

FIGURE 1.4   Applications of Deep Learning.

Computer Vision	Speech Analytics	Natural Language Processing
• Image Classification • Object Identification • Image Segmentation	• Automatic Speech Recognition • Speaker Diarization • Speaker Identification	• Machine Translation • Sentiment Analysis • Chat bot

FIGURE 1.5   Three Horizons of Deep Learning Applications.

- Natural Language Generation (NLG): Natural language generation is a course of converting structured code to unstructured data.

Some of the use cases of NLP are discussed in Figure 1.9, which can attract the researcher, and they are explained in Chapters 2 to 4.

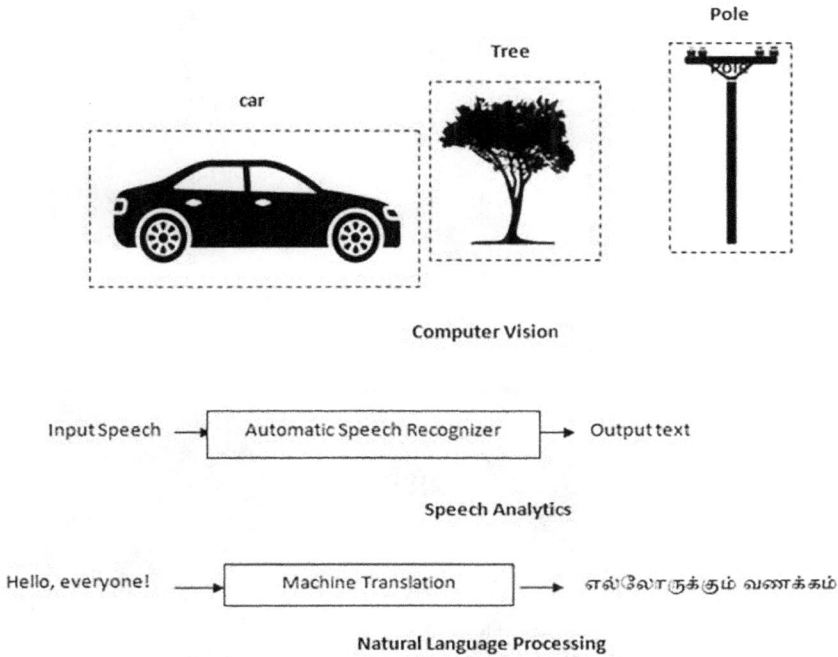

Computer Vision

Input Speech → Automatic Speech Recognizer → Output text

Speech Analytics

Hello, everyone! → Machine Translation → எல்லோருக்கும் வணக்கம்

Natural Language Processing

FIGURE 1.6   Deep Learning Applications.

1.1.4 SPEECH RECOGNITION

Speech is one of the most important forms of communication. Speech technologies are one of the most widely used applications in a wide range of domains. Automatic Speech Recognition (ASR) is a critical piece of technology for enabling and improving human–computer interactions. ASR systems translate spoken utterances into text. Examples of ASR systems are YouTube closed captioning, voicemail transcription, Siri, and Cortana. The acoustic model, language model, and pronunciation model are the three fundamental components of the ASR system, as shown in Figure 1.10.

- Acoustic Model: Identify the most probable phoneme given in the input audio.
- Pronunciation Model: Map phoneme to words.
- Language Model: Identify the likelihood of words in a sentence.

The different use cases of speech are depicted in Figure 1.11 and explained in the later part of the book in Chapter 5 to 7.

1.1.5 COMPUTER VISION

Human vision is both complex and beautiful. The vision system is made up of eyes that collect light, brain receptors that allow access to it, and a visual cortex that processes it. Computer vision AI research enables machines to comprehend and

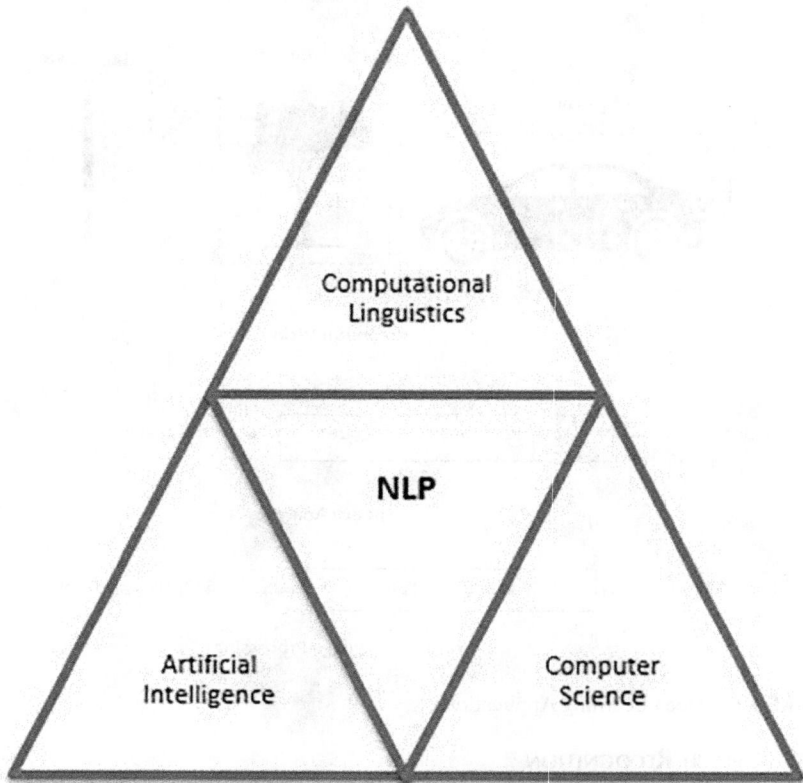

FIGURE 1.7   Natural Language Processing.

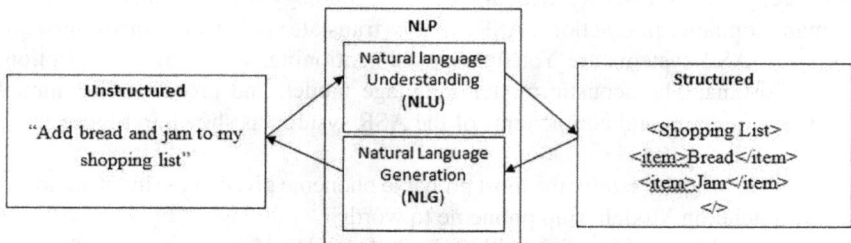

FIGURE 1.8   Variants of Natural Language Processing.

recognize the underlying structure of images, movies, and other visual inputs. The process of making a machine to understand and identify the underlying pattern behind an image is called computer vision. To prepare a calculation to sort out the example behind it and get a smart substance in the manner in which the human cerebrum does, we need to feed it incredibly huge data of millions of objects across thousands of angles. Some of the extensively used image-processing

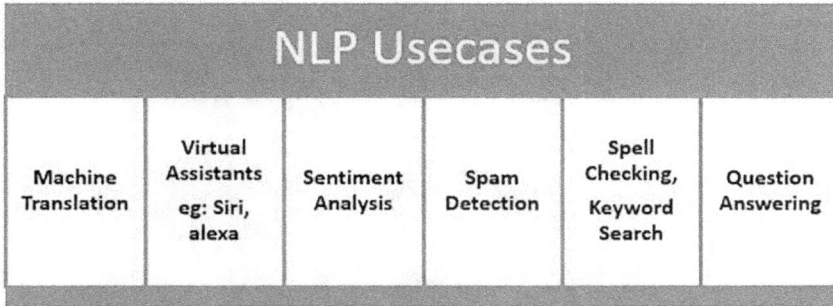

FIGURE 1.9   NLP Use Cases.

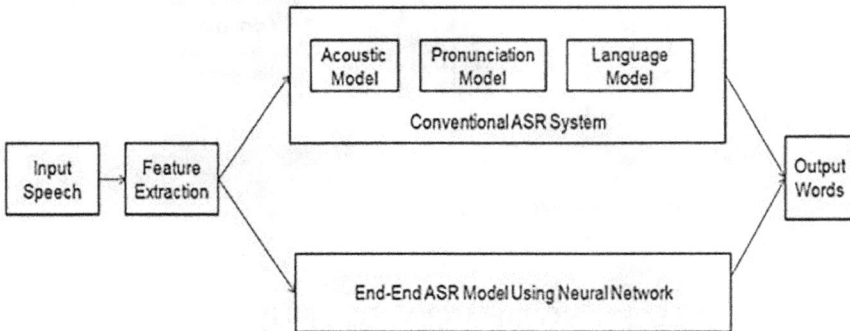

FIGURE 1.10   Automatic Speech Recognition.

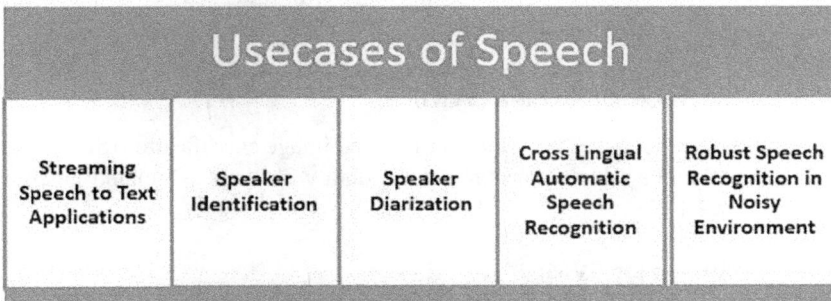

FIGURE 1.11   Use Cases of Speech.

algorithms are convolution neural network (CNN), recurrent neural network, and attention-based models. Figures 1.12 and 1.13 shows a sample computer-vision-based model and use cases of computer-vision-based applications which are explained in Chapters 8 to 10.

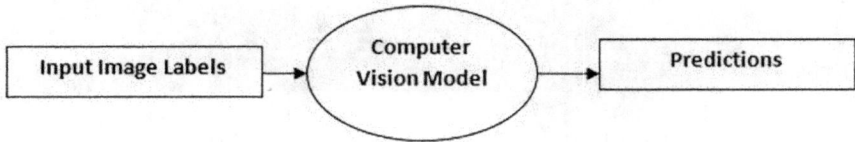

FIGURE 1.12 Computer Vision Model.

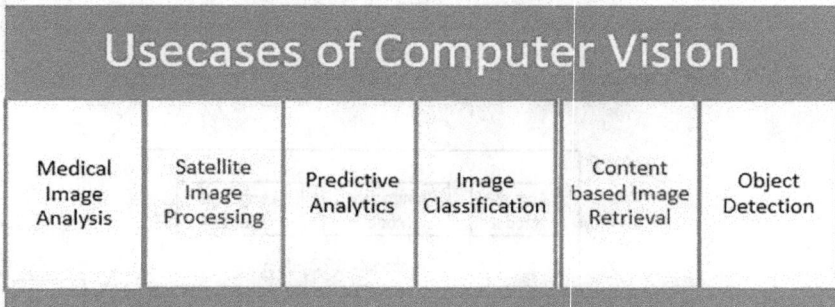

FIGURE 1.13 Use Cases of Computer Vision.

1.2 MACHINE LEARNING METHODS FOR NLP, COMPUTER VISION (CV), AND SPEECH

In this section, the machine learning algorithms that are utilized effectively for various applications of NLP, computer vision (CV), and speech analytics are discussed in detail.

1.2.1 Support Vector Machine (SVM)

SVM is a supervised learning technique for text and image classification. Supervised learning is a technique which learns about the data with their associated labels as shown in Figure 1.14.

Formal Definition for SVM:

SVM is a supervised algorithm that classifies cases into different groups.

The separator is otherwise known as hyper plane or decision boundary which is used to separate different classes. SVM is used to classify cases for both linearly and nonlinearly separable cases. Linearly separable is a one where a single line is sufficient enough to split two distinct classes whereas in a non-linearly separable case, the data points cannot be classified using a single separator. The examples for linearly and non-linearly separable cases are shown in Figure 1.15.

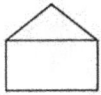

Labels: Spam/Not Spam

Input:
E-Mail Content

Spam E-Mail Detection

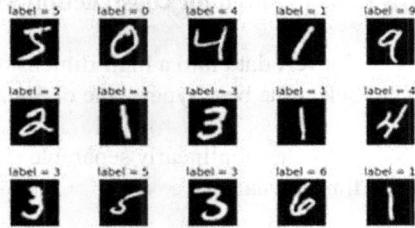

label = 5 label = 0 label = 4 label = 1 label = 9

label = 2 label = 1 label = 3 label = 1 label = 4

label = 3 label = 5 label = 3 label = 6 label = 1

Handwritten Digit Classification

FIGURE 1.14   Supervised Learning.

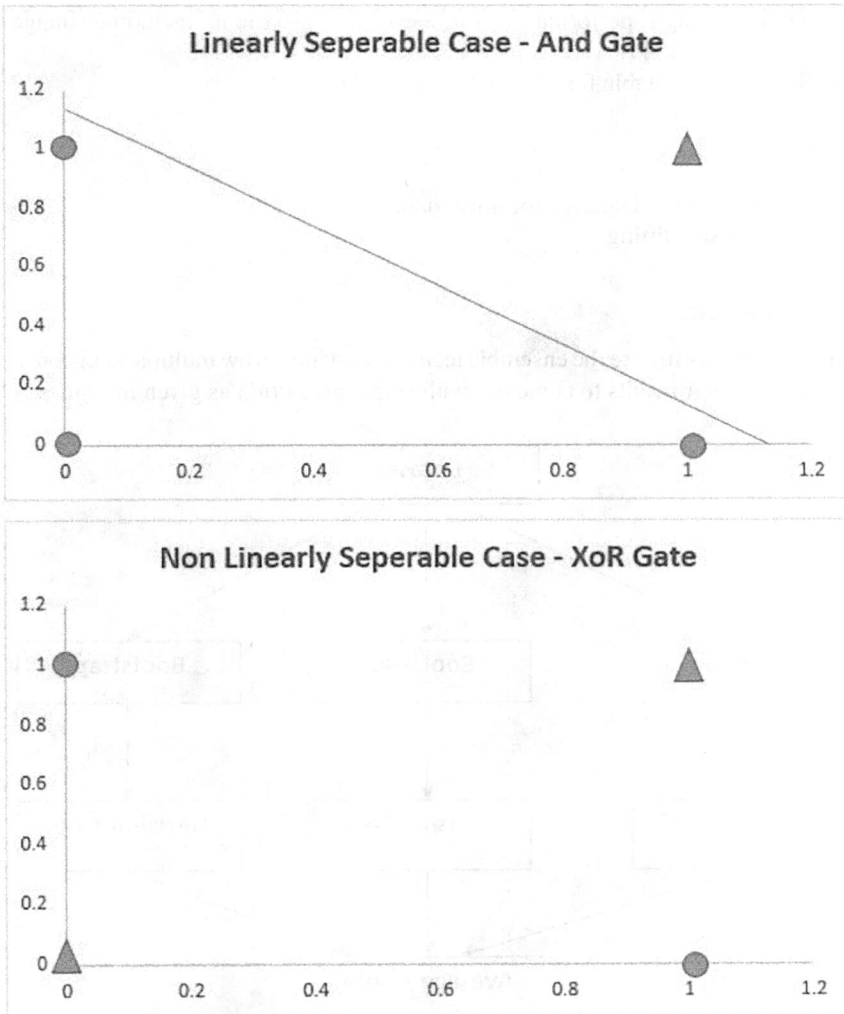

Linearly Seperable Case - And Gate

Non Linearly Seperable Case - XoR Gate

FIGURE 1.15   Linearly Separable versus Nonlinearly Separable Cases.

Steps to be followed for constructing SVM

1. Convert data into a high-dimensional feature space.
2. Select the best hyperplane or Separator.

SVM handles nonlinearly separable case using kernel tricks. It maps the data into higher dimensional space.

Pros:

- SVM is used to solve both linearly separable and nonlinearly separable problems
- Achieves high performance for various applications including image classification
- Simple and suitable for high-dimensional data

Cons:

- Performance is degraded for noisy data
- Prone to overfitting

1.2.2 BAGGING

Bagging and boosting are the ensemble techniques which grow multiple decision trees and combine their results to come out with single prediction as given in Figure 1.16.

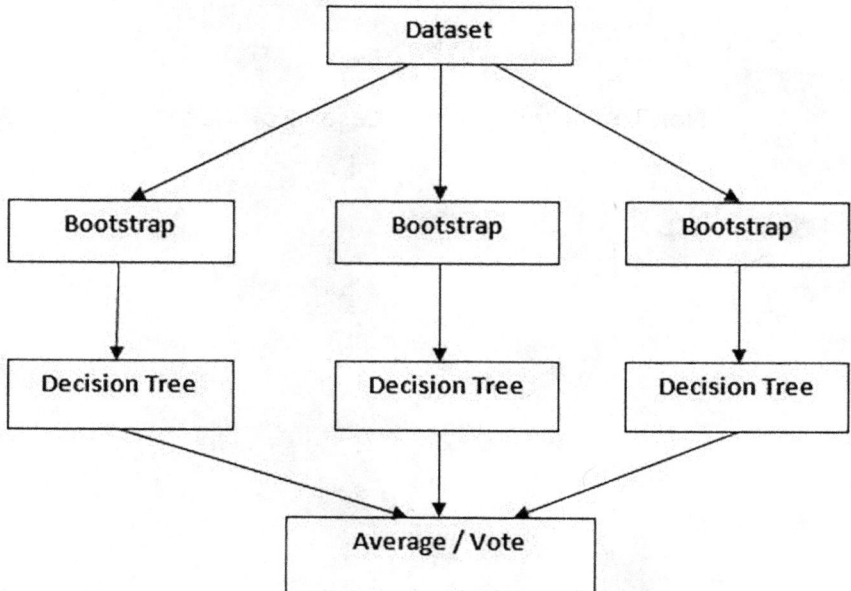

FIGURE 1.16 Random Forest Algorithm.

Random forest is an example of bagging mechanism with ensemble of trees which is widely used for various problems. In random forest, each tree is spitted on a random subset of variables, and trees are grown in a parallel fashion.

Pros:

- Excellent prediction performance on many problems
- Doesn't require extensive parameter tuning
- Easily parallelized across multiple CPUs

Cons:

- It is not a good choice for high-dimensional tasks such as text
- Difficult to interpret

1.2.3 GRADIENT-BOOSTED DECISION TREES (GBDTs)

Unlike the random forest to build the decision tree in a parallel fashion, as shown in Figure 1.17, the key idea of GBDT attempts to build a series of trees. In other words, the boosting algorithm grows trees sequentially, where each tree is trained and corrects the mistakes of the previous tree in the series. Sampling with replacement is carried out in boosting. Learning rate is the new parameter in GBDT, which is not used in random forest. The learning rate controls how each new tree is correcting the mistakes from the previous step. High learning rate indicates more complex trees, whereas low learning rate indicates a simpler tree. Like decision tress, GBDTs can handle various feature types.

1.2.4 NAIVE BAYES

Naive Bayes is a basic probabilistic model of how information in each class might have been created. They are also known as naive if the features are conditionally independent, given the class. Figure 1.18 depicts several flavors of the Naive Bayes classifier. They are defined next.

Bernoulli: As the name implies, it has binary properties. Since it permits us to address the presence or nonattendance of a word in the text, utilizing twofold qualities, Bernoulli naive bayes model is very useful for categorizing text documents. It does not, however, consider how frequently a word appears in the text.

Multinomial: This algorithm employs a set of count-based features that consider the frequency with which a feature, such as a word, appears in the document. Both Bernoulli and multinomial naive bayes are well suited for textual data.

Gaussian: A Gaussian naive bayes is used to calculate the mean and standard deviation of each class's component incentive. The classifier chooses the class that most intently mirrors the model information focus to be anticipated.

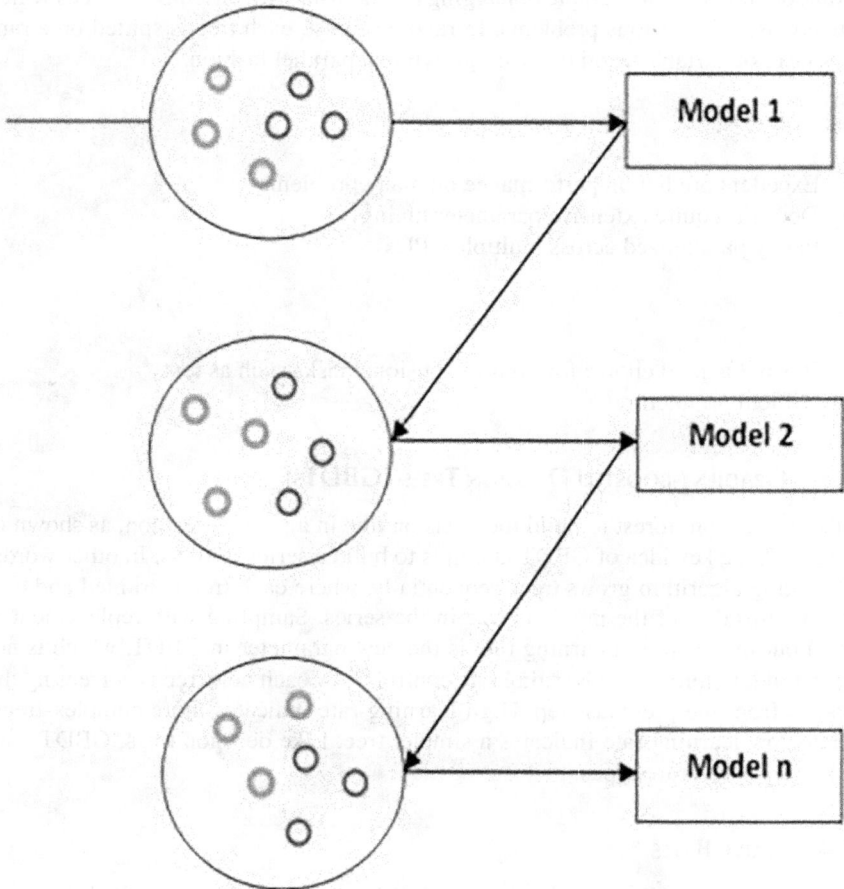

FIGURE 1.17   Boosting Model.

Pros:

- Succinct and efficient parameter estimation
- Works well with high dimensional data
- Easy to understand

Cons:

- The assumption that features are conditionally independent for the given class is not realistic
- Poor generalization performance

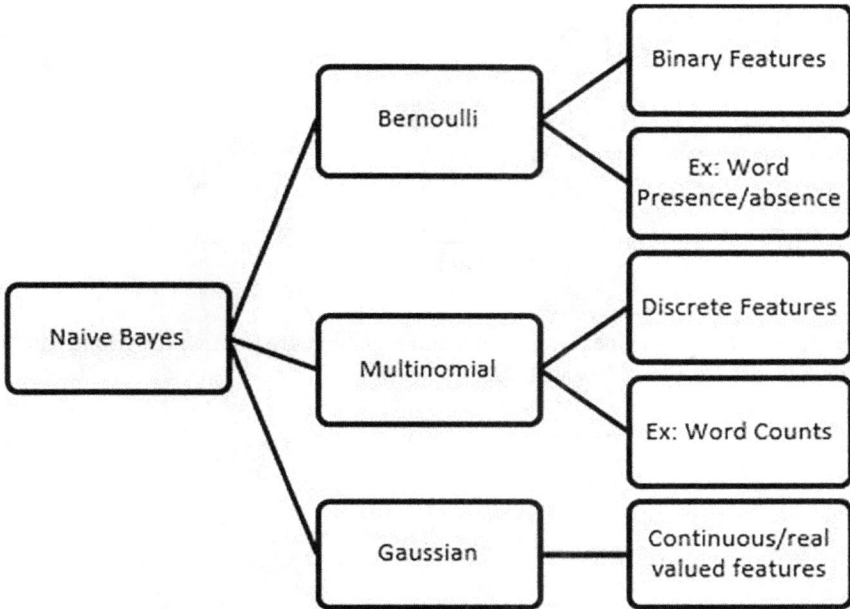

FIGURE 1.18 Variants of Naive Bayes.

1.2.5 LOGISTIC REGRESSION

Logistic regression is extensively used for binary classification. Most real-world situations have two outcomes: alive or dead, winner or loser, successful or unsuccessful—also known as binary, Bernoulli, or 0/1 outcomes.

Let us understand with a simple example: the main objective is to model the probability so that a player wins the match as given in Eq. (1.1). The probability value ranges between 0 and 1.

$$Pr(Win \mid Score, b_0, b_1) \tag{1.1}$$

The odds value is defined as the probability over one minus the probability. The value of odds ranges between 0 and ∞.

$$odds = \frac{Pr(Win \mid Score, b_0, b_1)}{1 - Pr(Win \mid Score, b_0, b_1)} \tag{1.2}$$

The log of the odds, otherwise called as logit, is usually the ranges between $-\infty$ and $+\infty$.

$$logit = log\left(\frac{Pr(Win \mid Score, b_0, b_1)}{1 - Pr(Win \mid Score, b_0, b_1)}\right) \tag{1.3}$$

The linear regression is given as

$$Win = b_0 \, Score + b_1 \tag{1.4}$$

The logistic regression equation is given as

$$\log\left(\frac{\Pr\left(Win|Score, b_0, b_1\right)}{1 - \Pr\left(Win|Score, b_0, b_1\right)}\right) = b_1 \, Score + b_0 \tag{1.5}$$

The generic forms for linear regression and logistic regression with sigmoid function is depicted in Figures 1.19 and 1.20.

$$y = logistic\left(b + w_1 x_1 + w_2 x_2 + w_n x_n\right) \tag{1.6}$$

$$= \frac{1}{1 + exp^{-\left(b + w_1 x_1 + w_2 x_2 + w_n x_n\right)}} \tag{1.7}$$

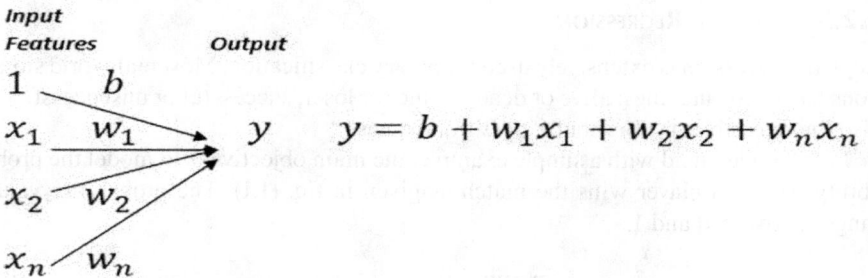

FIGURE 1.19 Linear Regression.

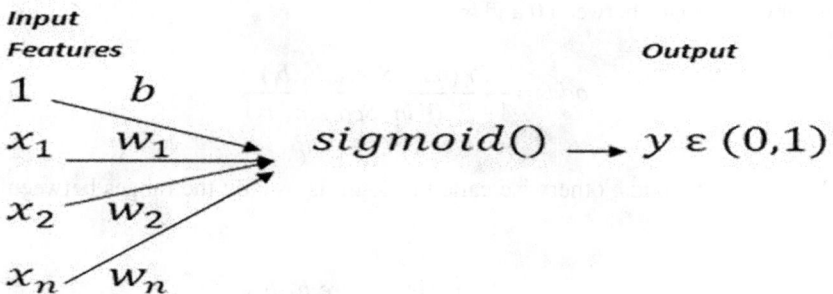

FIGURE 1.20 Logistic Regression with Sigmoid Function.

Pros:

- Widely accepted and proven technique for classification problems of health care domain
- Succinct to interpret and train for multiclass classifiers
- Logistic regression with regularization techniques can overcome overfitting problems

Cons:

- Less accurate for nonlinearly separable data
- Causes overfitting for higher dimensional feature set

1.2.6 DIMENSIONALITY REDUCTION TECHNIQUES

Dimensionality reduction is a feature selection technique. Lowering the high dimensional data into a more appropriate subset of related and useful features will improve the performance and accuracy of our model. The number of features or variables you have in your dataset determines the dimension of data. Suppose, if your dataset has two features or variables then it is a two-dimensional data. Similarly, dataset with three features or variables implies a three-dimensional data. Features capture the characteristics of data, so selecting the more useful features will have a great impact on our model outcome. The idea is to reduce the number of features to an optimal number. As the dimensionality increases, the complexity and the problem space that you are looking at increase exponentially as shown in Figure 1.21. In turn, as the space grows, the data becomes increasingly sparse.

The basic goal of dimensionality reduction is to find a smaller subset of dimensions or features as shown in Figure 1.22. The most commonly utilized dimensionality-decrease strategies are Principal Component Analysis, Singular Value Decomposition, and Latent Dirichlet Allocation.

Pros:

- Remove redundant features
- Remove multicollinearity
- Deal with the curse of dimensionality
- Identify the structure for supervised learning

Cons:

- Information loss

1.3 TOOLS, LIBRARIES, DATASETS, AND RESOURCES FOR THE PRACTITIONERS

The widely used machine learning tools and libraries are shown in Figures 1.23 and 1.24.

1 Dimensional Data with 10 positions

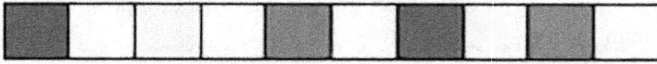

2 Dimensional Data with 100 positions

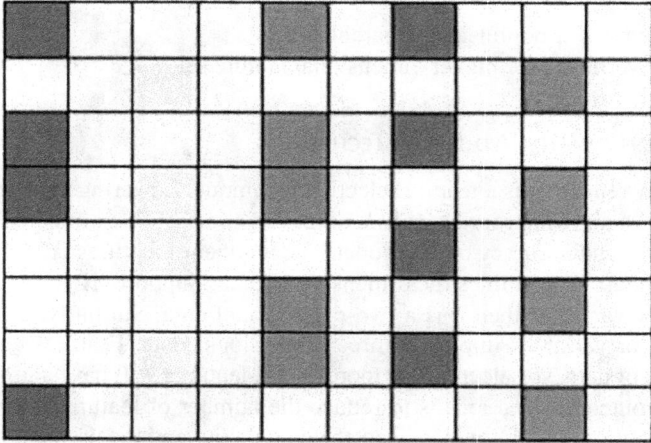

3 Dimensional Data with 1000 positions

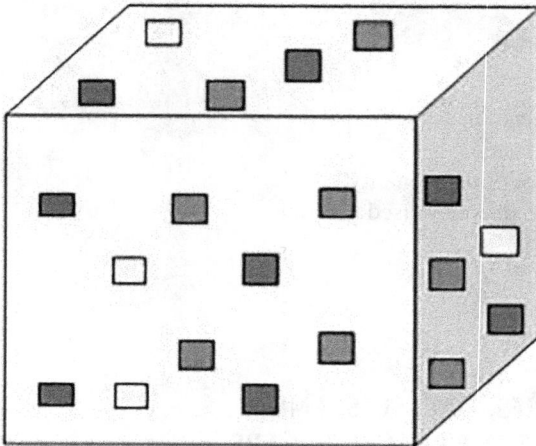

FIGURE 1.21 Curse of Dimensionality.

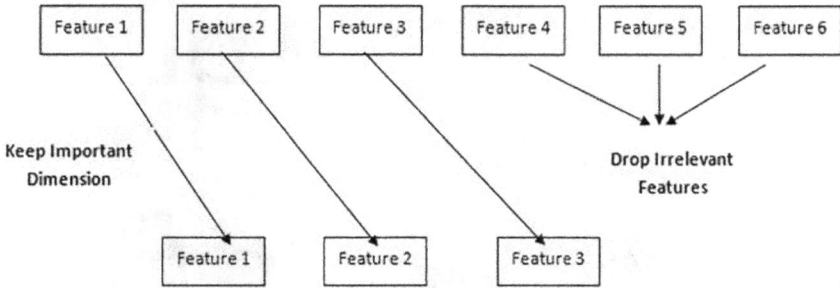

FIGURE 1.22 Dimensionality Reduction Techniques.

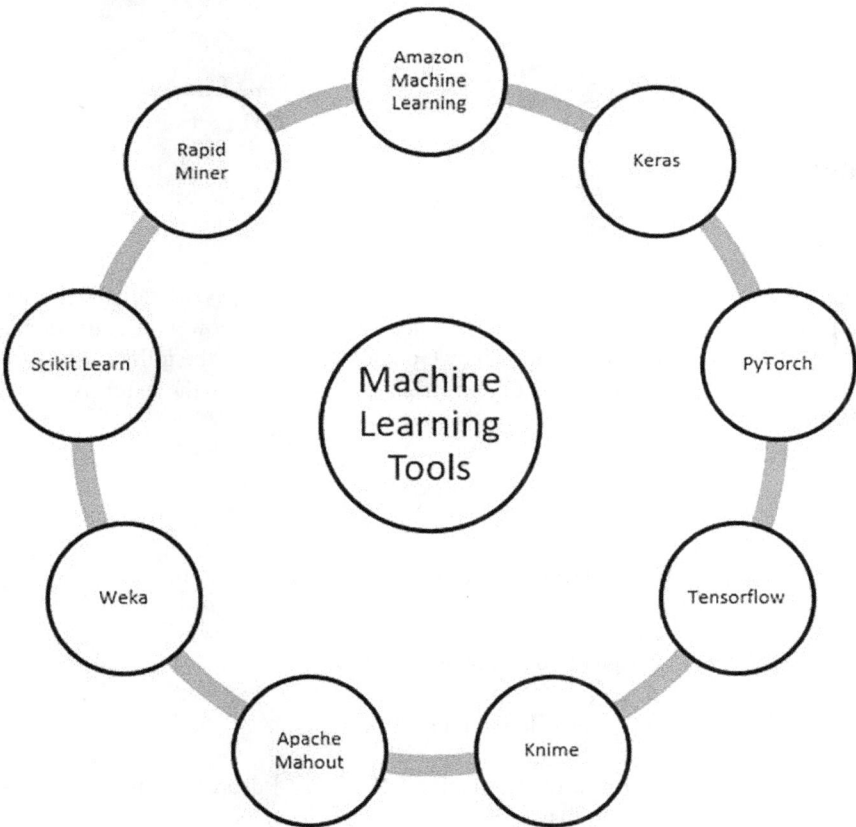

FIGURE 1.23 Machine Learning Tools.

FIGURE 1.24   Tools and Libraries.

1.3.1   TensorFlow

TensorFlow supports a wide range of solutions, including NLP, computer vision (CV), predictive ML, and reinforcement learning. TensorFlow is a start-to-finish open-source profound learning model by Google. You can use the following snippets to import TensorFlow libraries. Figures 1.23 and 1.24 depict the extensively used machine learning tools.
Import tensorflow as tf

```
print("TensorFlow version:", tf.__version__)
```

To import the inbuilt dataset from TensorFlow, use the following snippets:

```
import tensorflow_datasets as tfds
ds = tfds.load('mnist', split='train', shuffle_files=True)
```

Some of the features of TensorFlow are listed next.

- Enable the user to run their machine learning and deep learning codes on CPU and GPU platforms.
- Simple and flexible to train and deploy model in cloud.
- It provides eager execution that enables immediate iteration and intuitive debugging.

1.3.2 KERAS

Keras is an open-source API for developing sophisticated deep learning models. TensorFlow, a high-level interface, is used as the backend. It supports both CPU and GPU run time. It supports almost all the models of a neural network. Keras, being flexible, is simply suitable for innovative research.

1.3.3 DEEPLEARNING4J

Deeplearning4j is a deep learning library for Java virtual machines such as Scala and Kotlin. It can process massive amounts of data in a Java ecosystem, includes a deep learning framework with multiple threads as well as a single thread can be used in conjunction with Spark and Hadoop.

1.3.4 CAFFE

Caffe supports language like C, C++, Python, and MATLAB. It is known for its speedy execution. Caffe's model zoo consists of many pretrained models which can be used for various image-processing applications.

1.3.5 ONNX

The Open Neural Network Exchange, or ONNX, was created as an open-source deep learning ecosystem. ONNX, created by Microsoft and Facebook, is a deep neural network learning framework that allows developers to easily switch between platforms.

1.3.6 PYTORCH

PyTorch is an open-source machine learning tool developed by Facebook, which is extensively used in various applications like computer vision (CV) and NLP. Some of the popular use cases developed using PyTorch are hugging face transformers, PyTorch lightning, and Tesla Autopilot. Torchaudio, torchtext, and torchvision are parts of PyTorch. Use the code snippet given here and start working in PyTorch.

```
Import torch
    from torch import nn
    from torch.utils.data import DataLoader
    from torchvision import datasets
    from torchvision.transforms import ToTensor
```

PyTorch is a python library with a C++ core. Which empowers data scientist and AI engineers with profound deep learning models. It removes the cognitive overhead involved in building, training and deploying neural networks. It is built around the tensor class of multi dimensional arrays similar tonumpy. The following are the features of PyTorch:

- Very high performance on GPUs with CUDA
- Extensive neural network building block
- Built in backpropagation with autograd
- A simple, intuitive, and stable API

1.3.7 SCIKIT-LEARN

scikit-learn includes inbuilt datasets like iris, digits for classification, and the diabetes dataset for regression. The following are the simple code snippet for importing datasets from scikit-learn.

```
Pip install -U scikit-learn
    from sklearn import datasets
    iris = datasets.load_iris()
```

scikit-learn is a succinct and efficient tool for data mining and data analysis, built on NumPy, SciPy, and Matplotlib. It is extensively used for use cases in classification, regression, clustering, and dimensionality reduction.

1.3.8 NUMPY

NumPy stands for numerical Python and supports N-dimensional array objects that can be used for processing multidimensional data. NumPy supports vectorized operations.
NumPy is used in
- Mathematical and logical operations on arrays
- Fourier transforms
- Linear algebra operations
- Random number generation

NumPy is used to create an array.

```
Numpy.array(object)
NumPy is used to create an array with start, stop, and step value.
Numpy.arrange(start=1,stop=10,step=2)
```

Using NumPy, we can generate array with ones() and zeros() using the following snippets:

```
numpy.ones(shape,dtype)
    numpy.zeros(shape,dtype)
```

1.3.9 PANDAS

The pandas library gives elite execution, easy-to-use information structures, and logical capacities for the Python programming language. The open-source Python library gives superior execution information control and examination instruments by utilizing its strong

TABLE 1.1

Vector Representation

Python	Pandas Data Type	Description
int	int64	Numeric characters
float	float64	Numeric character with decimals

information structures. The pandas deals with data frames that are two-dimensional and mutable and store heterogeneous data. For importing necessary libraries, use the following snippet:

```
Import os
  import pandas as pd
```

The pandas library and Python use different names for data types as shown in Table 1.1.

1.3.10 NLTK

To work with human language data, Python scripts can be written using the open-source NLTK framework. It consists of text-processing libraries for NLP applications such as categorization, tokenization, stemming, tagging, parsing, and semantic reasoning. The following are the commands to use NLTK libraries:

```
import nltk
  nltk.download()
```

1.3.11 GENSIM

Gensim is a widely used tool for NLP applications, in particular for Word2Vec embedding. You can use the code given here to start with Gensim:

```
from gensim import corpora, models, similarities, downloader
Features of Gensim:
```

- Super fast: Highly optimized and parallelized C routines.
- Data streaming: Extensively used for data-streaming applications.
- Platform-independent: It supports Windows OS and Linux.
- Open source: It is available on Github.
- Ready-to-use pre-trained models are available.

Other NLP libraries are as follows:

- Stanford CoreNLP
- Apache OpenNLP
- Textblob Library
- IntelNLP Architect
- Spacy

1.3.12 DATASETS

Table 1.2 shows the widely used datasets in the cutting-edge research and industrial applications.

TABLE 1.2

Datasets

Natural Language Processing

Dataset Name	Application	Description
The Blog Authorship Corpus	Sentiment Analysis	The corpus contains 681,288 posts and north of 140 million words or roughly 35 posts and 7,250 words for each individual.
Amazon Product Dataset	Sentiment Analysis	Links (also viewed/also bought graphs), product metadata (descriptions, category information, price, brand, and image attributes), and product reviews are all included in this dataset (ratings, text, helpfulness votes).
Enron Email Dataset	Spam Email Classification	The corpus contains approximately 0.5 million messages in total.
Multi-Domain Sentiment Dataset	Sentiment Analysis	Amazon.com product reviews are included in the Multi-Domain Sentiment Dataset for a variety of product types.
Stanford Question Answering Dataset (SquAD)	Question and Answering Analysis	A brand-new dataset on reading comprehension that includes 10,000 queries submitted by Wikipedia crowd workers.
The WikiQA Corpus	Question and Answering	The WikiQA corpus is an open-source corpus with question and answer pair. Suitable for Question and Answering system.
Yelp Reviews	Text Classification	Open-source dataset with 6,990,280 reviews.
WordNet	Text Classification	Contains a large English lexical database. Cognitive synonyms (synsets) are groups of nouns, verbs, adjectives, and adverbs, and each conveys a unique idea.
Reuters Newswire Topic Classification	Text Classification	This dataset consists of 11,228 Reuters newswires organized into 46 themes.
Project Gutenberg	Language Modelling	28,752 English language books for research in language model.
Speech Analytics		
LibriSpeech	Speech to Text	Widely used Speech Recognition corpus in research. It consists of 1 k hours of read speech recorded in acoustic environment.
Acted Emotional Speech Dynamic Database	Emotion Recognition	Consists of speech samples for Greek language and is suitable for emotion analysis.
Audio MNIST	Speech to Text	The corpus consists of 30 k audio files of spoken digits (0–9) from multiple speakers.
VoxForge	Speech to Text	VoxForge is an open-source speech repository with varied dialects of the English language. Suitable for speech-to-text applications.

Natural Language Processing

Dataset Name	Application	Description
CREMA D: Crowd-Sourced Emotional Multimodal Actors	Emotion Recognition	The dataset consists of audio and visual data of speech under six different labels (happy, sad, anger, fear, disgust, and neutral). Suitable for multimodal speech emotion recognition research.
TED-LIUM	Speech to Text	452 hours of audio and 2,351 aligned automatic transcripts of TED Talks.
Common Voice	Speech to Text	Common Voice is designed by Mozilla to create a free database for speech recognition software. It consists of audio files of different dialects.
Multilingual LibriSpeech	Speech to Text for regional languages	Consists of audio files of Indian languages for ASR applications.
IEMOCAP	Emotion Recognition	Multimodal database with audio, video, and text modality, which contains approximately 12 hours of data. Suitable for multimodal learning applications and effective computing research.

Computer Vision

Dataset Name	Application	Description
CIFAR-10	Image Classification	The CIFAR-10 dataset consists of 60 k high-resolution colour images with ten class labels.
Image Net	Image Classification	The Image Net project is a large-scale image repository designed for computer vision applications and research with approximately around 14 million images.
MS COCO	Object Detection	MS COCO (Microsoft Common Objects in Context) is a large-image dataset that contains 328,000 images of people and common objects.
MNIST	Digit Classification	One of the extensively used image datasets for computer vision applications with approximately around 70,000 images.
IMDB Wiki Dataset	Object Identification	Open-source dataset of face images with gender and age labels for training.
American Sign Language Letters Dataset	Sign language recognition	Includes sign language postures for letters A to Z.
Alzheimer's Disease Neuroimaging Initiative (ADNI)	Medical Image Classification	Includes scanned brain images of normal and Alzheimer-affected patients.
SpaceNet	Geospatial Image Analysis	The SpaceNet Dataset is an open-source dataset with millions of high-resolution road map and building images for geospatial machine learning research.

1.4   SUMMARY

It is fairly obvious that the advent of deep learning has triggered several of the practical use cases in industrial applications. This chapter gives readers a high-level overview of AI and its subsets like machine learning and deep learning. It also gives a gentle introduction on machine learning algorithms like SVM, Random Forest, Naive Bayes, and vector representation of NLP. The detailed discussion on tools, libraries, datasets, and frameworks will enable readers to explore their ideas.

This book is organized as follows: Chapters 2, 3, and 4 will describe the fundamental concepts and major architectures of deep learning utilized in NLP applications and aid readers in identifying the suitable NLP-related architectures for solving many real-world use cases. Chapters 5, 6, and 7 will detail the concepts of speech analytics and the idea behind sequence-to-sequence problems with various pretrained architecture, and also discusses the cutting-edge research happening in this domain. Chapters 8, 9, and 10 walk through the basic concepts behind CV with a focus on real-world applications and a guide to the architecture of widely used pretrained models.

BIBLIOGRAPHY

Aurélien Géron, *Hands-On Machine Learning with Scikit-Leaᵐ, Keras, and TensorFlow*, 2nd Edition, O'Reilly Media, Inc, 2019
Aurélien Géron, *Neural Networks and Deep Learning*, O'Reilly Media, Inc, 2018
Ben Lorica, Mike Loukides, *What Is Artificial Intelligence?*, O'Reilly Media, Inc, 2016
Christopher M. Bishop, *Pattern Recognition and Machine Learning,* Springer, 2006
David Beyer, *Artificial Intelligence and Machine Learning in Industry*, O'Reilly Media, Inc, 2017
Ian Goodfellow and Yoshua Bengio and Aaron Courville, *Deep Learning*, MIT Press, 2016
S Lovelyn Rose, L Ashok Kumar, D Karthika Renuka, *Deep Learning Using Python*, Wiley India, 2019
S Sumathi, Suresh Rajappa, L Ashok Kumar, Surekha Paneerselvam, *Machine Learning for Decision Sciences with Case Studies in Python*, CRC Press, 2022
Tom Mitchell, *Machine Learning*, McGraw Hill, 1997
Valliappa Lakshmanan, Martin Görner, Ryan Gillard, *Practical Machine Learning for Computer Vision*, O'Reilly Media, Inc, 2021

2 Natural Language Processing

LEARNING OUTCOMES

After reading this chapter, you will be able to:

- Understand NLP, NLP Pipeline, and Text Pre-Processing tasks.
- Understand the basic concepts behind Text Pre-Processing with word-embedding techniques.
- Understand the concepts on language models with evaluation techniques.

2.1 NATURAL LANGUAGE PROCESSING

A subfield of computer science, AI, and computational linguistics called "Natural language processing" (NLP) deals with computers and human (natural) languages. Its objective is to make it possible for computers to converse with people naturally, i.e., utilizing language rather than just a collection of symbols. Speech synthesis, document retrieval, understanding natural language, machine translation, and information retrieval are all tasks that fall under the umbrella of NLP. NLP, or natural language processing, is the study of computers and other devices that can comprehend human language. In other words, computers can understand the instruction by humans, draw inferences, and decide on the basis of the words (also known as sentences) they hear. In essence, it is the capacity of computers and other technologies to converse with a language they understand. NLP is already utilized in many different applications, such as machine translation and chatbots, and it will continue to expand and create new uses. At its most fundamental level, NLP encompasses functions like statistical machine translation, document classification, spelling checks, and optical character recognition. In more advanced forms, it might entail a machine that can converse with a human in natural language and reply in response to input. Consider an instance in which you can communicate with a computer in a language that humans can understand or a setting where a car may hear and respond to voice commands or a setting where someone communicate with a friend and they could both understand one another. It makes it possible for computers to decode and examine spoken language. Additionally, it aids with human-like language processing in machines. It is a very complicated topic since it involves many elements that machines are unfamiliar with. These components include context, pauses, and intonation. These machines' algorithms are specifically created to comprehend these components. These algorithms combine NLP, machine learning, and deep learning. As a result, a system that comprehends the meaning of words and sentences has been created. The phrase "natural language processing" refers to various activities including the computer processing of human language.

DOI: 10.1201/9781003348689-2

One of the fastest-growing disciplines of computer science, natural language processing (NLP) is increasingly permeating all facets of the tech industry. The discipline of NLP is extremely complex, but it is expanding quickly. Despite being a decades-old subject of study, NLP appears to be gaining ground due to the development of new technologies and the rising use of AI in daily life. To improve lives by making machines more human, many of these new technologies use NLP in some way, shape, or form. A simple statement like "I want to eat sushi in San Francisco" can be understood by Google's search engine, for instance, and will produce pertinent results. NLP is the underlying technology of this. This is accomplished by the computer doing tasks including word, sentence, and voice part recognition. Customer service, chatbots, automated email responses, and numerous other business and technological fields all make an extensive use of NLP. The field of customer service has been greatly benefited by NLP. Imagine being able to instruct a chatbot to add more pepperoni and sausage when placing an online pizza order. Natural Language Processing has a somewhat uncertain future, although it is obvious that it will become more integrated into our daily lives over the next few years. Understanding natural human language is the process of NLP. In other terms, it refers to a system's capacity to comprehend human language. When people hear about NLP, the first thing they think of is the iPhone personal assistant "Siri." Siri is able to understand what you are saying, but it cannot interpret intentions. The goal of NLP is to create machines that can comprehend and comprehend human language in general.

2.2   GENERIC NLP PIPELINE

The entire data is converted into strings via NLP over text or video, and the primary string then goes through several phases (the process called processing pipeline). Depending on voice tone or sentence length, it employs trained pipelines to supervise your input data and recreate the entire string. The component returns to the main string after each pipeline. It then moves on to the following element. The components, their models, and training all affect the capabilities and efficacy. Data capture, text cleaning, pre-processing, feature engineering, modeling, evaluation, deployment, monitoring, and model update shown in Figure 2.1 are the essential steps in a generic pipeline for creating modern, data-driven NLP systems.

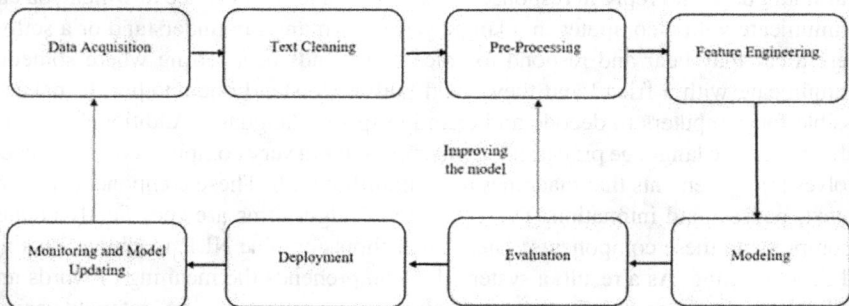

FIGURE 2.1   NLP Pipeline.

2.2.1 DATA ACQUISITION

The NLP pipeline heavily relies on data. Because of this, it is crucial to gather the necessary information for NLP projects. It is readily available, but sometimes more work is required to gather these priceless data.

a) Scrape web pages

To develop a program that can sum up the most significant news in 100 words or less, data from websites and webpages pertaining to current events must be scraped for this.

b) Data augmentation

With the help of a number of NLP techniques, it is possible to expand an existing tiny dataset. These techniques, which are also known as data augmentation, aim to make use of linguistic traits to produce text that is syntactically comparable to the underlying text data. Although they can seem like hacks, they actually function extremely effectively.

c) Back translation

Consider the Spanish sentence s1 as an example. After being translated into the target language, in this case English, it becomes sentence s2. The sentence that was first translated as "s2" was changed to "s3" in Spanish. s3 was finally added to the dataset after it was discovered that s1 and s3 have extremely similar meanings with a few minor differences.

d) Replacing entities

The entity names will be swapped out with those of other entities to create new datasets. For instance, I would like to visit Australia. Just rename Australia to something else, like Antarctica.

e) Synonym replacement

Pick any words in a sentence, for example, "s" words at random inside a sentence, which are not stop words that substitute their use for these terms. Usually, stop words are eliminated from texts before they are processed for analysis.

f) Bigram flipping

A bigram or diagram is a sequence of two adjacent tokens from a string of tokens, typically letters, syllables, or words. When n = 2, a bigram is an n-gram. The frequency analysis of each bigram in a string is frequently used in many applications, including computational linguistics, encryption, and speech recognition, for simple statistical text analysis. For example, "I'm going to watch a movie." Here, we swap out the bigram "going to" for the inverted version, "to Going."

2.2.2 TEXT CLEANING

After data collection, it is important that the data be presented in a way that the computer can understand. This fact should be taken into account when the model is being trained, as the text may contain various symbols and words that have no apparent significance. Eliminating words from the text before feeding them into the model is effective in text cleaning. This technique is also called as Data Cleaning. Some text cleaning procedures are given next.

a) HTML tag cleaning

There is the possibility to scrape through different web pages when gathering the data. Beautiful Soup and Scrapy offer a variety of tools for parsing web pages. As a result, there are no HTML tags in the content that was gathered.

b) Unicode normalization

It may come across other Unicode characters, such as symbols, emoji, and other graphic characters, which may be encountered, when cleaning the data. Unicode normalization is used to parse such non-textual symbols and special characters. This indicates that in order to store the text in a computer, it must be transformed into some sort of binary representation. Text encoding is the name of this procedure. Code snippets given next explains a simple word that was given with an emoji; by using the Unicode normalization, it will deliver the respected emoji-encoded formats. Table 2.1 represents the emojis with their respective Unicode.

c) Spelling correction

The information may contain mistakes due to hurried typing, the usage of shorthand, or slangs seen on social media sites like Twitter. It is necessary to treat these data before feeding them to the model because using them cannot not improve the model's ability to forecast. There won't be a reliable way to repair this, but we may still make good efforts to lessen the problem. For possible spell checking in Python, Microsoft released a REST API.

TABLE 2.1
Unicode Normalization for Emojis

Emoji	Unicode
☺	U+263A
~	U+1F603
♥	U+2665
♠	U+2660
✓	U+2713

d) System-specific error correction

Different PDF documents are encoded differently; therefore, occasionally, it might not be possible to extract the entire text or the content might have an incorrect structure. Many libraries exist to extract text from PDF documents, although they are far from being ideal, including PyPDF, PDFMiner, and others. Scanned documents are another typical source of text data. Optical character recognition, or OCR, is frequently used to extract text from scanned documents, and libraries like Tesseract are often used.

2.3   TEXT PRE-PROCESSING

Pre-processing text simply refers to cleaning the text and analyzing the text for the specific activity. A task combines an approach with a domain. Using the TF-IDF (approach) to extract the top keywords from Tweets is an illustration of a task (domain).

Task = approach + domain

Stop words must be removed during the pre-processing stage because they are used in another activity and will likely cause the situation of missing certain common words. In NLP software, the following typical pre-processing steps are used:

- Sentence segmentation and word tokenization are preliminary steps.
- Steps that are frequently taken include stop word removal, stemming and lemmatization, deleting numbers and punctuation, lowercasing, etc.
- Advanced processing, including co-reference resolution, POS labelling, and parsing.

2.3.1 NOISE REMOVAL

Unwanted information can occasionally be found in the source of text data. For instance, text data that has been extracted from a webpage will have HTML tags. Additionally, it must remove punctuation, spaces, numerals, special characters, etc., from the text data. Pre-processing includes the crucial and domain-specific step of removing noise from text data. Depending on the particular NLP assignment you're working on, the notion of noise will vary. Punctuation might be considered noise in some situations, but crucial information in others. For instance, you could want to eliminate punctuation while estimating a Topic Model. However, punctuation must always be left in the data while training a transformer.

2.3.2 STEMMING

The technique of stemming involves words returning with inflection to their original form. In this instance, the "root" can simply be a canonical version of the original

TABLE 2.2

Stemmer Stemming

	Original Word	Stemmed Words
0	connect	connect
1	connected	connect
2	connects	connect
3	connections	connect

word and not a true root word. By chopping off the ends of words, the rudimentary heuristic technique of stemming attempts to appropriately convert words into their root form. Therefore, since the ends were simply chopped off, the terms "Connect", "Connected", and then "Connects" can be changed to "Connect" instead of using the original words; refer to Table 2.2 for better understanding. For stemming, various algorithms exist. The Porters Method is the most often used algorithm and is well known to be empirically successful for English. Here is a Porter stemming demonstration in Table 2.2 for practice. Stemming is helpful for standardizing terminology and addressing concerns with sparsity and to reveal documents that mention "Machine Learning classes" in addition to those that mention "Machine Learning class." To find the most pertinent papers, every word variation should be matched. In comparison to better constructed features and text enrichment techniques like word embeddings, stemming provides a slight improvement for classification accuracy.

2.3.3   TOKENIZATION

Text must be converted into a collection of fundamental components, referred to as Tokens, in order for a machine to simulate natural language. Tokenization is the process of dividing text data into smaller or more fundamental components. Tokenizers are programmers that separate text into tokens, which are small, distinct entities. There are no formal guidelines regarding how to tokenize a string; however, it is quite typical for a tokenizer to employ spaces to tokenize a string (e.g., separate it into individual words). Some tokenizers divide words into their constituent parts rather than just concentrating on white space. After tokenizing a document or group of documents, you can determine the set of distinctive tokens that frequently appear in your data. The term "vocabulary" refers to this collection of distinctive tokens. As a result, the content of your vocabulary is directly impacted by your tokenizer. Vocabulary will contain individual words if tokens are single words; however, if tokens are sub-word strings or characters, vocabulary will be made up of those elements rather than individual words. Once vocabulary has been established, it can map each distinct token to a distinct integer value. In NLP, words are changed into numbers in this manner.

```
from nltk.tokenize import word_
tokenize tokentext = "Knowledge is Power."
tokenized = word_tokenize (tokentext)
print (tokenized)
```

White Space Tokenization is the simplest method of tokenization, where tokenization is carried out simply separating input anytime a white space is observed. Although this method is quick, it has the limitation that it only functions for languages like English where there is white space between phrases. Other tokenization technologies include hugging face tokenizers, tokenization with NLTK (Natural Language ToolKit), spaCy tokenizers, etc.

2.3.4 LEMMATIZATION

Similar to stemming, lemmatization aims to take a word out of its inflections and transfer it to its basic form. The sole distinction is that lemmatization attempts to carry out the process correctly. It actually changes words to their true roots rather than simply cutting them off. It may make use of WordNet-style dictionaries or other specialized rule-based approaches for mappings; better, for example, would be comparable to "good."

When it comes to search and text classification, lemmatization does not seem to be any better than stemming. Depending on the technique used, it can be significantly slower than using a very simple stemmer and may need comprehension of the word's part of speech in order to generate the proper lemma. The accuracy of text categorization using neural architectures is unaffected by lemmatization. The additional expense could or might not be appropriate. An example of a lemmatized term formed from the original word is shown in Table 2.3.

2.3.5 STOP WORD REMOVAL

Stop words are a collection of words that are regularly used in a language. English stop terms include "a", "the", "is", "are," and others. By eliminating low information terms from text, stop words are used with this logic. For instance, if the search question is "what is text pre-processing?", it tells a search engine to give documents a lower priority than things that discuss text pre-processing when used with a search engine. Stop words are often employed in search engine algorithms, topic modelling, topic extraction, and text categorization software.

Despite its success in search and topic extraction systems, it has been discovered that the removal of stop words is not essential in categorization systems. To make room for more fair model sizing, it does help to reduce the number of letters in words that are taken into account. Here is an example of stop word removal in action. A dummy letter, W, is used to substitute all stop words:

TABLE 2.3
Lemmatization for Original Words

	Original _Word	Lemmatized Word
0	trouble	trouble
1	troubling	trouble
2	troubled	trouble

Original sentence = this is a book full of information and we need to study

Sentence with stop words removed = W W W book full W information W W W W study

Stop word lists can be made specifically for domains or drawn from pre-existing sets. Some libraries (like Sklearn) allow you to delete words that appeared in X% of your texts, which may also have the effect of reducing word usage.

2.3.6 PARTS OF SPEECH TAGGING

The technique of text for a particular piece of a speech is marked up using POS tagging (parts of speech tagging) based on its context and significance. It is in charge of reading texts in a language and assigning a specific token to each word (parts of speech). It is also known as grammatical tagging. The words in a noun, adjective, verb, or adverb are marked with POS tags. Table 2.4 explain the POS with their tags for 4 POS for understandings.

2.4 FEATURE ENGINEERING

Feature engineering is the collective of techniques that will carry out the task of feeding the pre-processed text. It is also known as feature extraction. The objective of feature engineering is to convert the text's qualities into a numeric vector that the machine learning algorithms can comprehend. There are two distinct methods used in practice for feature engineering.

Classical NLP and traditional ML pipeline

Any ML pipeline must include the stage of feature engineering. The raw data is transformed into a machine-consumable format by feature engineering processes. In the traditional ML pipeline, these transformation functions are often created and tailored to the task at hand. Consider the task of emotion classification for e-commerce product reviews. Counting the positive and negative words in each review is one technique to translate the reviews into meaningful "numbers" that might assist forecast their moods (positive or negative). If a feature is helpful for a task or not, it can be determined statistically. One benefit of using handcrafted features is that the model can still be understood because it is feasible to measure precisely how much each feature affects the model prediction.

TABLE 2.4
Part of Speech with Tags

Part of Speech (POS)	Tag
Noun	n
Verb	v
Adjective	a
Adverb	r

- **DL pipeline**

In this case, after pre-processing, the raw data is given straight to a model in the DL pipeline. The model has the ability to "learn" characteristics from the data. As a result, these features are better suited to the job at hand and generally produce better performance. However, the model becomes less interpretable because all of these features are learned through model parameters.

2.5 MODELING

The next step is to figure out on developing a practical solution from NLP Pipeline. The development of the model will depend on the techniques and the guideline information due to their less information present in the text. To raise the performance of the model with addition in data, some ways should be followed that are listed next.

2.5.1 START WITH SIMPLE HEURISTICS

ML may not have a significant impact on a model's development at first. A dearth of data may be a contributing factor in some cases, but human-made heuristics can also be a fantastic place to start. For instance, we might maintain a blacklist of domains that are only used to send spam for email spam-classification duties. Emails from certain domains can be filtered using this information. A blacklist of email terms that indicate a high likelihood of spam might also be used for this classification, in a similar manner. Using regular expression is a popular method for including heuristics in the system. Tokens Regex from Stanford NLP and rule-based matching from spaCy are two tools that are helpful for creating complex regular expressions to gather additional data.

2.5.2 BUILDING YOUR MODEL

A set of straightforward heuristics is a fine place to start, but as the system develops, adding ever-more-complicated heuristics may lead to a rule-based system that is complex. It might be challenging to administer such a system and even more challenging to identify the root of issues to identify system that will be less difficult to maintain as it advances. Additionally, as more data is gathered, ML models begin to outperform simple heuristics. Heuristics are frequently used in conjunction with the ML model at that moment, either directly or indirectly. There are basically two methods for doing that:

- *Create a feature from the heuristic for your ML model*

It is ideal to utilize these heuristics as features to train your ML model when there are numerous heuristics whose individual behavior is deterministic but whose combined behavior is unclear in terms of how it predicts. It may enhance the ML model in the email spam classification example by including features like the number of words from the blacklist in a particular email or the email bounce rate.

- *Pre-process your input to the ML model*

It is advantageous to use a heuristic first, before feeding data into a machine learning model, if it has a really high prediction rate for a specific class. For instance, it is preferable to categorize an email as spam rather than sending it to an ML model if particular terms in the email have a 99% likelihood of being spam.

2.5.3   METRICS TO BUILD MODEL

The final phase is to develop the model, which provides good performance and is also production-ready, after following the baseline of modelling, starting with straightforward heuristics and moving on to building model, in need to perform numerous iterations of the model-building procedure. Here are some problems with the methods:

- **Ensemble and stacking**

Instead of using a single model, it is typical to employ a variety of ML models that often address various parts of the prediction problem. There are two ways to accomplish this: either using the output of one model as the input for another or sequentially using one model after another to arrive at the final output. Model stacking is the term for this. It can also combine forecasts from other models to come up with a final projection. The term for this is model ensemble.

- **Better feature engineering**

Performance may improve as a result of better feature engineering. For instance, feature selection is used to discover a better model when there are several features.

- **Transfer learning**

A model that has been trained for one task is repurposed for a different, related task using the machine learning technique known as transfer learning. BERT can be used, as an illustration, to fine-tune the email dataset for the classification of email spam.

- **Reapplying heuristics**

At the conclusion of the modeling pipeline, it is feasible to go back and look at these examples again to identify any patterns in the defects and utilize heuristics to fix them. The model predictions can be improved by using domain-specific knowledge that is not automatically captured in the data.

2.6   EVALUATION

NLP model can be evaluated with their performance after building the model. The evaluation metrics depends on the NLP tasks. During the model building stage and deployment stages, the evaluation metrics are used. The metrics will suffice in the processing, and it decides the model performance. There are some evaluation metrics available for NLP task, and they are listed here:

- **BLEU**

Understudy, commonly known as BLEU, is a precision-based metric that assesses the correctness of machine-translated text from one natural language to another by computing the n-gram overlapping between the reference and the hypothesis. The BLEU is the ratio of overlapping n-grams to all n-grams in the hypothesis. More specifically, the numerator holds the sum of the overlapping n-grams across all hypotheses, while the denominator contains the total number of n-grams across all hypotheses.

- **NIST**

NIST weights each matched n-gram in accordance with its information gain in addition to BLEU (Entropy or Gini Index). Over the set of references, the information gain for an n-gram composed of the letters w1, . . ., wn is determined. It was built on top of the BLEU metric with a few changes. While BLEU analyzes n-gram accuracy by giving each one the same weight, NIST additionally assesses how informative a certain n-gram is. It is intended to give more credit when a paired n-gram is unusual and less credit when it is common in order to decrease the chance of manipulating the measure by producing useless n-grams.

- **METEOR**

The drawbacks of BLEU include the fact that it only supports accurate n-gram matching and ignores recollection. To fix these issues, METEOR (Metric for Evaluation of Translation with Explicit Ordering) was developed. It uses lax matching criteria and is based on the F-measure. METEOR takes into account a matched unigram even if it is equivalent to a unigram in the reference but does not have an exact surface level match.

- **ROUGE**

BLEU is based on precision, whereas ROUGE (Recall-Oriented Understudy for Gisting Evaluation) is based on recall. Four different versions of the ROUGE metric exist: ROUGE-N, ROUGE-L, ROUGE-W, and ROUGE-S. Similar to BLEU-N, ROUGE-N measures the number of n-gram matches seen between reference and the hypothesis. These metrics are used to assess machine translation and automatic summarization software in NLP. The measurements contrast an autonomously generated summary or translation with a reference summary or translation that was created by a human.

- **CIDEr**

According to the concept of CIDEr (Consensus-based Image Description Evaluation), each image should have several reference captions. It is predicated on the idea that an image's reference captions will frequently contain n-grams related to the image. Each n-gram in a sentence is given a weight by CIDEr using TF-IDF on the basis of how frequently it appears in the corpus and the reference set for that specific circumstance (term-frequency and inverse-document-frequency). But since they are less likely to be instructive or pertinent, n-grams that often exist throughout the dataset (e.g., in the reference captions of different photos) are given a lower weight using the inverse-document-frequency (IDF) term.

- **SPICE**

SPICE (Semantic Propositional Image Caption Evaluation) is another technique for captioning photos that focus on n-gram similarity; however, in this algorithm, the semantic propositions implied by the text are more weight. SPICE communicates semantic propositional material through scene-graphs. The SPICE score is calculated by comparing the scene-graph tuples of the recommended sentence to all of the reference phrases. Then, scene graphs are created from the hypothesis and references. WordNet synonyms for METEOR are taken into consideration by SPICE when matching tuples.

- **BERT**

The use of BERT to obtain word embeddings demonstrates that outcomes from contextual embeddings and a simple averaged recall-based metric are equivalent. The greatest cosine similarity in between embeddings of any reference token and any token in the hypotheses is used to calculate the BERT score.

- **MOVERscore**

The ideal matching measure developed by MOVERscore, which determines the Euclidean distances between words or n-grams using contextualized embeddings, is based on the WMD metric. MOVERscore uses soft/partial alignments to provide many-to-one matching, similar to how WMD uses Word2Vec embeddings to support partial matching, in contrast to BERT score, which allows one-to-one hard word matching. Four NLG tasks—machine translation, image captioning, abstractive summarization, and text generation—have been shown to have competitive correlations with human evaluations.

2.7 DEPLOYMENT

Deployment is one of the stages in an NLP pipeline's post-modeling phase. Once it is satisfied by the model's performance, it is ready to be deployed in production, where in connect with NLP module to the incoming data stream, the output is usable by downstream applications. An NLP module is typically deployed as a web service, and it is critical that the deployed module is scalable under high loads.

2.8   MONITORING AND MODEL UPDATING

After the deployment, the model's performance is continuously tracked. As it is instructed to make sure that the outputs generated by the developed models on a daily basis make sense, monitoring for NLP projects and models must be approached differently than it would be for a typical engineering project. The model will iterate depending on fresh data once it has been installed and has begun collecting new data in order to keep up with forecasts.

2.9   VECTOR REPRESENTATION FOR NLP

A variety of word-embedding approaches are used when creating representations of words to capture the meaning, semantic relationship, and context of various words. Word embedding is a technique for producing dense vector representations of words that contain specific context phrases. These are improved variants of straightforward bag-of-words models, which are frequently used to represent sparse vectors and include word counts and frequency counters. Word embeddings use a method to train fixed-length dense and continuous-valued vectors on a large text corpus. Every word functions as a learned and relocated point in a semantically related vector space that surrounds the target word. When words are represented in vector space, a projection is made that groups words with similar meanings together.

2.9.1   ONE HOT VECTOR ENCODING

It is used to transform the nominal character into a column vector representation. To understand the math behind it, let us consider the same example from the aforementioned vocabulary which consists of 1,000 words. For the word "enjoy," the one hot representation is shown in Figure 2.2.

2.9.2   WORD EMBEDDINGS

In NLP, text processing is a technique used to tidy up text and get it ready for model creation. It is adaptable and contains noise in many different ways, including emotions, punctuation, and text written in specific character or number formats. Starting with text processing, several Python modules make this process simpler and offer a lot of versatility thanks to their clear, simple syntax. The first is NLTK, which stands for "natural language toolkit," which is helpful for a variety of tasks like lemmatizing, tokenizing, and POS. There isn't a single statement that isn't a contraction, which means we frequently use terms like "didn't" instead of "did not." When these words are tokenized, they take on the form "didn't," which has nothing to do with the original word. There is a collection of terms called contractions that deals with such words.

BeautifulSoup is a package used for online scraping, which acquire data containing HTML tags and URLs, so BeautifulSoup is used to handle this. Additionally, we are utilizing the inflect library to translate numbers into words.

Words	Integers
a	1
able	2
about	3
enjoy	520
Zebra	1000

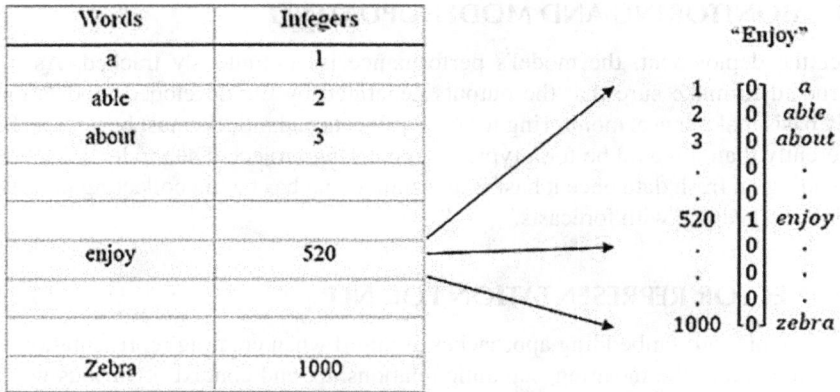

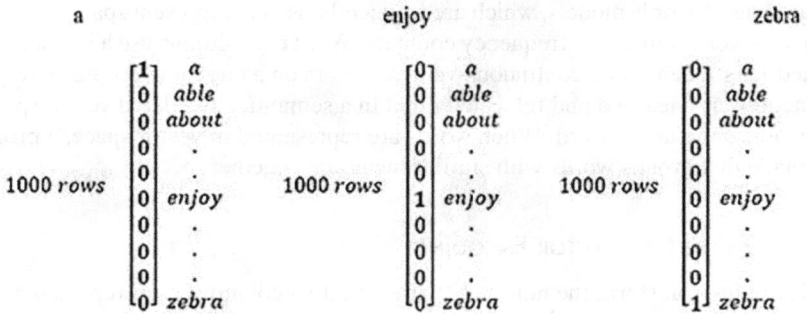

FIGURE 2.2 Sample of One Hot Vector Encoding.

Word embedding is a technique used in NLP to represent words for text analysis. This representation frequently takes the form of a real-valued vector that encodes the definition of the word, presuming that words that are close to one another in the vector space would have related meanings. Using word embeddings, which are a type of word representation, it is possible to show words with similar meanings. They are a dispersed representations for the text and may be one of the major technological advancements that allows deep learning algorithms to excel at difficult NLP challenges. Each word is represented as a real-valued vector in a predetermined vector space during a process known as word embedding. The method is sometimes referred interpreted as "deep learning" since each word is given to a unique vector, and the vector values are learned similar to a neural network. The distributed representation is learned via word usage. This enables words that are frequently used to naturally have representations that accurately convey their meaning.

Neural network embeddings serve three main objectives:

- Finding the embedding space's closest neighbors. Based on user interests or cluster categories, they can be utilized to provide suggestions.
- Finding supervised task's input for a machine learning model.
- Used for the purpose of categorization and concept visualization.

TABLE 2.5

Bag of Words Calculation

	amazing	an	anatomy	best	great	greys	is	series	so	the	TV
0	1	1	1	0	0	1	1	1	0	0	1
1	0	0	1	1	0	1	1	1	0	1	1
2	0	0	1	0	1	1	1	0	1	0	0

2.9.3 BAG OF WORDS

Natural language processing employs the text modelling technique known as "bag of words." To explain it formally, it is a method for feature extraction from text data. This approach makes it simple and flexible to extract features from documents. A textual example of word recurrence in a document is called a "bag of words" since any information about the placement or organization of the words within the document is discarded. Table 2.5 explains the bag of words concept with the calculation of bag-of-words approach:

> Line 1: *Grey's Anatomy* is an amazing TV series!
> Line 2: *Grey's Anatomy* is the best TV series!
> Line 3: *Grey's Anatomy* is so great

2.9.4 TF-IDF

Utilizing the statistical method TF-IDF (term frequency-inverse document frequency), anybody can determine how relevant a word is to each document in a group of documents. To accomplish this, the frequency of a word within a document and its inverse document frequency across a collection of documents are multiplied. It has a wide range of uses, with automated text analysis being the most important, including word scoring in machine learning algorithms for NLP. TF-IDF was created for document search and information retrieval. It works by escalating in accordance with how frequently a term appears in a document but is balanced by how many papers contain the word.

TF-IDF is calculated for each word in a document by multiplying two separate metrics:

- The number of times a word appears in a text. The simplest way to calculate this frequency is to simply count the instances of each word in the text. Other ways to change the frequency include the document's length or the frequency of the term that appears most frequently.
- The phrase "inverse document frequency" appears frequently in a group of documents. This relates to how frequently or infrequently a word appears in all writings. The nearer to 0 a word is, the more common it is. This metric can be calculated by taking the entire collection of papers, then dividing it by the total number of documents including a word, and then computing the logarithm.

- This value will therefore be close to 0 if the word is widely used and appears in numerous papers. If not, it will go close to 1.

The result of multiplying these two figures is the word TF-IDF score in a document. The more relevant a word is in a given document, the higher the score.

The following formula is used to determine the TF-IDF score for the word t in a document d from a document set D:

$$tf\ idf\ (t, d, D) = tf\ (t, d). idf\ (t, D)$$

where

$$tf\ (t, d) = \log\ (1 + freq\ (t, d))$$

2.9.5 N-GRAM

N-grams are continuous word, symbol, or token combinations in a document. They are the adjacent groups of items in a document. N-grams are relevant when performing NLP activities on text data. A list of the different categories for n-grams is shown in Table 2.6, where n is the positive integer value that includes the total number of n-grams, and in the term it describes the different categories for n-grams.

Examples for n-gram representation in text:

- "Candy"—Unigram (1-gram)
- "Candy Crush"—Bigram (2-gram)
- "Candy Crush Saga"—Trigram (3-gram)

The sample given here demonstrates the variety of n-gram types seen in typical literature. The N number of grams was determined by counting the sequence of words that were present in the circumstance where each repetition of the phrases was treated as a separate gram.

2.9.6 WORD2VEC

The Word2Vec (W2V) technique extracts a vector representation of each word from a text corpus. In order to represent the distribution of words in a corpus C, the Word2Vec model mixes many models. The Word2Vec approach uses a neural

TABLE 2.6
N-gram Categories

n	Term
1	Unigram
2	Bigram
3	Trigram
n	n-gram

TABLE 2.7

Word2Vec Representation

	I	Like	enjoy	machine	learning	vector	studying	.
I	0	2	1	0	0	0	0	0
like	2	0	1	0	1	0	0	0
enjoy	1	0	0	0	0	0	1	0
machine	0	1	0	0	1	0	0	0
learning	0	0	0	1	0	0	0	1
vector	0	1	0	0	0	0	0	1
studying	0	0	1	0	0	0	0	1
.	0	0	0	0	1	1	1	0

network model to learn word associations from a vast corpus of text. Once trained, a model of this kind can identify terms that are synonyms or can recommend new words to complete a sentence. Word2Vec, as the name suggests, uses a vector of specified numbers to represent each unique word. The vectors are selected with care so that a straightforward mathematical function (cosine similarity between vectors) may be used to determine how semantically similar the words represented by each vector are to one another. Table 2.7 gives a straightforward illustration showing how words are represented as vectors.

Line 1: I enjoy studying.
Line 2: I like vector.
Line 3: I like machine learning.

2.9.7 GLOVE

The use of embeddings rather than other text representation approaches like one hot vector encodes, TF-IDF, and bag-of-words has produced numerous impressive outcomes on deep neural networks with difficulties like neural machine translations. Additionally, some word-embedding techniques, such GloVe and Word2Vec, may eventually achieve performance levels comparable to those of neural networks. GloVe is an abbreviation for Global Vectors in word representation. It is an unsupervised learning system created by Stanford University that tries to create word embeddings by combining global word co-occurrence matrices from a corpus. The main principle of GloVe word embeddings is to utilize statistics to determine the relationship between the words. The co-occurrence matrix, as opposed to the occurrence matrix, tells how frequently a particular word pair appears with another. In the co-occurrence matrix, each value reflects a pair of words that frequently appear together.

2.9.8 ELMO

The numerous aspects of word use, including as syntax and semantics, as well as the ways in which these uses vary across linguistic contexts, are modeled by the deep contextualized word representation known as ELMo (i.e., to model polysemy). These word vectors are learned functions of the internal states of a deep bidirectional language model that has been pre-trained on a substantial text corpus (biLM). They greatly enhance the state-of-the-art for many challenging NLP problems, including sentiment analysis, question answering, and textual entailment, and are easy to include into already-existing models. ELMo shows the context in which each word is spoken throughout the entire dialog. Since ELMo is character-based, the network may infer the vocabulary tokens from training by using morphological cues.

Examples for ELMo are as follows:

1. I love to watch movies.
2. I am wearing a watch.

Here, the word "watch" is used as a verb in the first phrase and as a noun in the second. These words are referred to as polysemous words since their context varies between sentences. This type of word nature can be handled by ELMo with greater success than GLOVE or FastText.

2.10   LANGUAGE MODELING WITH N-GRAMS

The simplest and most obvious use of n-gram modeling in NLP is word prediction, which is frequently used in text messaging and email. The n-gram model is frequently the sole ML technology, and we need to do this task because the entire use case depends around anticipating what will happen next on the basis of what has already happened. This is definitely not the case with many other applications of n-grams, which depend on a variety of technologies combining to create a single coherent engine. For instance, n-gram models are also used by auto-correct systems, such as the ones in word processors that fix grammar and spelling. It makes sense to use an n-gram model to determine whether to employ the words "there," "their," or "they're" in a phrase on the basis of the context in which they appear. However, applying grammatical rules necessitates considering sentences as a whole, so what comes next could matter just as much as what came before.

This is still a pattern identification issue that ML tools like Deep Learning can handle. In this approach, the computer "gets a concept of where to seek" and can complete its duty more quickly and effectively than if it had to search through a big lexicon. n-grams are simply one small component of a larger tech stack used in speech recognition activities that are even more complicated, such as voice assistants and automated transcription. An acoustic model will examine the sound waves that make up speech and turn them into phonemes, the fundamental components of speech, before we even get to the n-gram. Then, a different kind of language model will convert those sounds into words and the individual characters that make up those words. In this situation, n-grams frequently play a supporting role by reducing the search space. In essence, the n-gram will examine the words that have already been transcribed in order to limit the options for the subsequent word.

2.10.1   Evaluating Language Models

2.10.1.1   Perplexity

When normalized by the number of words in the test set, perplexity is the multiplicative inverse of the probability that the language model is assigned to the test set. A language model is more accurate if it can anticipate terms not included in the test set, or if P (a phrase from a test set) is higher. Therefore, better language models will yield test sets with lower perplexity levels or higher probability values. A more accurate language model would place words according to conditional probability values supplied using the training set, creating meaningful sentences. Therefore, the state that the perplexity value given to the language model is based on a test set asserts how effectively the language model can anticipate the next word and thereby build a meaningful sentence.

2.10.1.2   Shannon Visualization Method

This technique uses the trained language model to produce sentences. If the trained language model is a bigram, then the following sentences are produced by the Shannon Visualization Method:

- Based on its probability, select a bigram at random (s, w). Decide on a random bigram (w, x) based on its probability, and so on till we reach a decision. Then join the words together. The punctuation marks s> and /s> here denote the beginning and conclusion of the phrases, respectively.

2.10.2   Smoothing

In order for all plausible word sequences to occur with some probability, a language model's inferred probability distribution must be flattened (or smoothed). This frequently entails enlarging the distribution by shifting weight from regions of high likelihood to regions of zero probability. Smoothing works to increase the model's overall accuracy in addition to preventing zero probability. Parameter estimation (MLE) should be employed in a linguistic model using training data. Due to the likelihood that both test sets would contain words and n-grams with a chance of 0, we are unable to evaluate our MLE models using unseen test data. All probability mass is assigned to events in the training corpus through relative frequency estimation.

Examples for smoothing are:

Training data: The goat is an animal.
Test data: The dog is an animal.
If we use the unigram model to train,
P(the) = count(the)/(Total number of words in the training set) = 1/5. Likewise, P(goat) = P(is) = P(an) = P(animal) = 1/5
To evaluate (test) the unigram model,
P(the goat is an animal) = P(the) * P(goat) * P(is) * P(an) * P(animal) = 0.00032
It becomes zero when applying the unigram model to the test data since P(dog) = 0.
Dog was never mentioned in the training data.

2.10.3 KNESER-NEY SMOOTHING

The main purpose of the Kneser-Ney smoothing technique is to determine the probability distribution of n-grams in a document based on their histories. Due to its absolute discounting technique, which omits n-grams with lower frequencies by deducting a fixed amount from the probability's lower order terms, it is regarded as the most effective form of smoothing. The answer is to "smooth" language models, which shifts some probability in the direction of unidentified n-grams. There are numerous approaches, but interpolated, modified Kneser-Ney smoothing yields the best results. The main purpose of the Kneser-Ney smoothing approach is to determine the probability distribution of the n-grams in a document on the basis of their histories. Explaining Kneser-Ney smoothing, language models are an important component of NLP and are used in anything from machine translation to spell checking. A language model can determine whether a given text belongs to a specific language, given any random piece of text. It will offer a memory-efficient, high-throughput, and simple-to-use library by building the main functionality in C++ and making it accessible to R via Rcpp, using cmscu as the querying engine to implement the computationally difficult modified Kneser-Ney n-gram-smoothing technique.

2.11 VECTOR SEMANTICS AND EMBEDDINGS

2.11.1 LEXICAL SEMANTICS

The lexical meaning of a word (or lexeme) is its sense (or meaning) as it is given in a dictionary, commonly known as the underlying meaning, denotative meaning, or semantic meaning, unlike what the grammatical connotation would suggest (or structural meaning). The branch of linguistics that focuses on the analysis of lexical meaning is known as lexical semantics. There is no necessity that a word's lexical and structural meanings match up. When we consider the word "cat," which has an object as its structural and lexical meaning, these meanings are consistent. Consequently, a word's structural and lexical meanings frequently have opposing or even surprising effects on one another. For instance, "protection" refers to a thing in its structural meaning, although its lexical meaning for the word refers to "process," and "cage." It refers to an object in its structural meaning, whereas its lexical meaning refers to "process" (the term "process" describes a specific situation as an instance here). A crucial aspect of the link between structural and lexical meanings is how lexical meanings impose restrictions on grammatical rules. To state the laws of grammar, however, we must remove ourselves from the lexical constraints imposed on the grammatical conventions of diverse languages. Different languages' lexical constraints on grammar rules cannot be used to convey grammar rules.

2.11.2 VECTOR SEMANTICS

A word is represented by vector semantics in a multidimensional vector space. Words are embedded in a certain vector space; vector models are also known as embedding. In NLP, the vector model has several benefits. In sentimental analysis, for instance, it establishes a boundary class and determines if the sentiment is positive or negative (a binomial classification). The ability of vector semantics to spontaneously learn from text without complicated labeling or supervision is another significant practical benefit. These benefits have led to the vector semantics status as the industry standard for NLP applications like sentiment analysis, named entity recognition (NER), and topic modeling. While semantic analysis concentrates on larger chunks, lexical analysis is centered on smaller tokens.

2.11.3 Cosine for Measuring Similarity

The text similarity test must assess how closely two texts are related in terms of context or meaning. There are several text similarity metrics, including Jaccard similarity, cosine similarity, and Euclidean distance. Each of these metrics has a unique specification that measures how similar two queries are to one another. Cosine Similarity is one of the metrics used in NLP to compare the text in two documents, regardless of their size. A word is visualized as a vector. The text documents are visualized as vector objects in an n-dimensional space. The cosine of the angle formed by two n-dimensional vectors projected in a multi-dimensional space is measured mathematically by the cosine similarity metric. The range of the cosine similarity between two documents is 0 to 1.

Formula for calculating Cosine Similarity:

$$\text{Similarity} = \cos(\theta) = \frac{A.B}{\|A\| \|B\|} = \frac{\sum_{i=1}^{n} A_i B_i}{\sqrt{\sum_{i=1}^{n} A_i^2} \sqrt{\sum_{i=1}^{n} B_i^2}}$$

The orientation of two vectors is the same if the cosine similarity score is 1. The closer the value is to 0, the less similar the two documents are. The cosine similarity metric is preferable to Euclidean distance because there is still a potential that two texts that are far away by Euclidean distance will be similar to one another in terms of context.

2.11.4 Bias and Embeddings

In a wide range of tasks, including language translation, pathology, and playing video games, neural network models are capable of spotting patterns and illuminating structure. Moreover, harmful biases can appear in neural networks and other machine learning models in a variety of different ways. For instance, face categorization algorithms may not function as well for women of color, and African Americans may experience higher rates of transcribing mistakes than White Americans. For instance, TensorFlow Hub has opened up access to its platform, and developers now have free access to a large number of machine learning models that have already been trained. Developers must be mindful of the biases these models contain and how those biases could manifest in their applications if they are to use them effectively. In human data, biases are automatically encoded. Knowing this is an excellent place to start, and managing it is currently being discussed. Because Google is committed to developing products that are helpful for everyone, they are continually looking into analysis and mitigation techniques for unintended bias.

2.12 SUMMARY

The theories of NLP, NLP Pipeline, and various text pre-processing methods, including noise removal, stemming, tokenization, lemmatization, stop word removal, and parts of speech tagging, are all covered in detail with examples in this chapter. The concepts of NLP in language models for n-grams were also explained, and the text semantics of vector, lexical, cosine, and bias were also covered in detail.

BIBLIOGRAPHY

Farha, I. A., & Magdy, W. (2021, April). Benchmarking transformer-based language models for Arabic sentiment and sarcasm detection. In *Proceedings of the sixth Arabic natural language processing workshop* (pp. 21–31). Toronto: Association for Computational Linguistics.

Feldman, J., Lakoff, G., Bailey, D., Narayanan, S., Regier, T., & Stolcke, A. (1996). L 0—The first five years of an automated language acquisition project. In *Integration of natural language and vision processing* (pp. 205–231). Dordrecht: Springer.

Lewis, P., Ott, M., Du, J., & Stoyanov, V. (2020). Pretrained language models for biomedical and clinical tasks: Understanding and extending the state-of-the-art. In *Proceedings of the 3rd clinical natural language processing workshop, Online* (pp. 146–157). Toronto: Association for Computational Linguistics, Anna Rumshisky, Kirk Roberts, Steven Bethard, Tristan Naumann (Editors).

Liddy, E. D. (2001). Natural language processing. In *Encyclopedia of library and information science*, 2nd ed. New York: Marcel Decker, Inc.

Maulud, D. H., Zeebaree, S. R., Jacksi, K., Sadeeq, M. A. M., & Sharif, K. H. (2021). State of art for semantic analysis of natural language processing. *Qubahan Academic Journal*, *1*(2), 21–28.

Raina, V., & Krishnamurthy, S. (2022). Natural language processing. In *Building an effective data science practice* (pp. 63–73). Berkeley, CA: Apress.

Smelyakov, K., Karachevtsev, D., Kulemza, D., Samoilenko, Y., Patlan, O., & Chupryna, A. (2020). Effectiveness of preprocessing algorithms for natural language processing applications. In *2020 IEEE international conference on problems of infocommunications. Science and technology (PIC S&T)* (pp. 187–191). New York: IEEE.

Spilker, J., Klaner, M., & Görz, G. (2000). *Processing self-corrections in a speech-to-speech system*, edited by Wolfgang Wahlster. Verbmobil: Foundations of Speech-to-Speech Translation.

Sun, F., Belatreche, A., Coleman, S., McGinnity, T. M., & Li, Y. (2014, March). Pre-processing online financial text for sentiment classification: A natural language processing approach. In *2014 IEEE conference on computational intelligence for financial engineering & economics (CIFEr)* (pp. 122–129). New York: IEEE.

Wasim, M., Asim, M. N., Ghani, M. U., Rehman, Z. U., Rho, S., & Mehmood, I. (2019). Lexical paraphrasing and pseudo relevance feedback for biomedical document retrieval. *Multimedia Tools and Applications*, *78*(21), 29681–29712.

3 State-of-the-Art Natural Language Processing

LEARNING OUTCOMES

After reading this chapter, you will be able to:

- Understand the basic sequence-to-sequence modeling task.
- Identify the basic building block of RNN and attention.
- Identify the various language models such as BERT and GPT3.
- Apply deep-learning-based sequential models for NLP applications.

3.1 INTRODUCTION

It is a huge challenge for the computers to understand information as the way we do, but advances in technology are helping in bridging the gap. Technologies like ASR, NLP, and CV are helpful in transforming in a more useful way than ever before. The ability of computers to understand and interpret the human languages was envisaged by Alan Turing in 1950 as a hallmark of computational intelligence. Many commercial applications of today make use of NLP models for their downstream modules. A wide variety of applications such as search engine and chatbots utilize AI-based NLP in their back end. If you observe, most of the NLP problems are aligned toward sequential data. In general, human thoughts have persistence. For example, you comprehend each word in this chapter on the basis of how you comprehended the words preceding it. A conventional convolution neural network is not suitable for such complex time series data as it accepts predetermined input vector such as image and produces predetermined output vector such as class labels. To overcome the shortcomings of CNN, sequential models such as RNN comes into the picture. A recurrent neural network (RNN) is a kind of brain network that is utilized to tackle consecutive information like text, sound, and video. RNN is able to store information about the previous data using memory, and they are great learners of sequential data. One of the major disadvantages of RNN is that it is unable to capture contextual information when the input sequence is larger. To tackle the long-term sequential dependency, Long Short Term Memory (LSTM) is introduced, and it is evident that it is suitable for many real-world NLP problems. Ian Goodfellow came up with an idea of attention mechanism which is yet another cutting-edge model which handles sequential NLP data in a more contextual way by applying its focus to input sequential data.

This chapter explains the various sequential models such as RNN, LSTM, attention and transformer-based models for NLP applications. These models are widely used for various NLP use cases as follows:

DOI: 10.1201/9781003348689-3   **49**

- Machine translation

Machine translation is used to automatically translate text in different languages without the assistance of human linguists.

- Sentiment analysis

Sentiment analysis, often known as opinion mining, is a technique used in natural language processing (NLP) to determine the emotional context of a document.

- Chatbot

Rather than offering direct contact with a genuine human specialist, a chatbot is a product program that mechanizes communications and is utilized to lead online discussions through text or text-to-discourse.

- Question-answering system

Building systems that respond automatically to questions asked by people in natural language is the focus of question-answering system.

- Name-entity recognition

Recognizing and sorting named elements referred to in unstructured text into pre-laid out classes like individual names, associations, areas, and so forth is the goal of the name-entity recognition.

- Predictive text

Through word suggestions, predictive text is an input method that makes it easier for users to type on mobile devices.

3.2   SEQUENCE-TO-SEQUENCE MODELS

Sequence modeling represents sequential data such as speech and natural language. Sequential models utilize sequential data pair with the sequence of discrete outputs. Before understanding sequence modeling in detail, let us see what sequence and sequence labeling are about.

3.2.1   SEQUENCE

The word sequence means continuous flow of data with respect to time. In general, sequence consists of multiple data points where data points depend on each other in a complicated way. For instance, sequence could be something like sentence, medical EEG signal, and speech wave form as given in Figure 3.1.

3.2.2   SEQUENCE LABELING

Sequence labeling is the process of identifying suitable sequence of labels for the input sequential data. A well-known example of sequence labeling is parts of speech

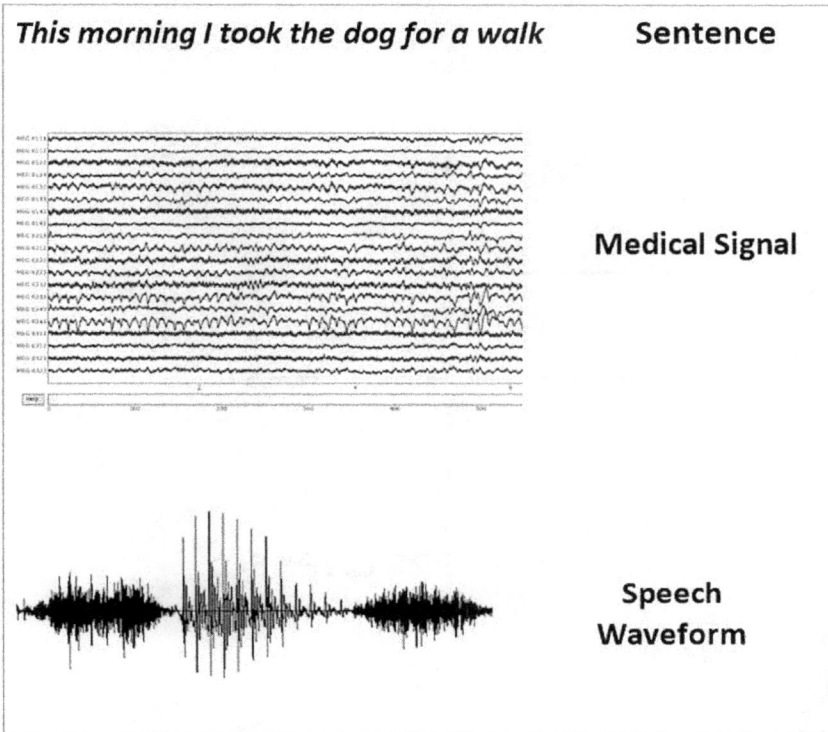

FIGURE 3.1 Samples for Sequence-to-Sequence Modeling.

tagging. For in Figure 3.2, the input text sequence is mapped to output parts of speech tag using neural network models.

3.2.3 SEQUENCE MODELING

Sequence modeling is a process of capturing the contextual information by analyzing the input sequence to predict the better outcome. It requires a memory to store the relevant information. For instance, a sample sentence "This morning I took the dog for a walk" is made up of several interconnected terms. The goal of sequence modelling, as shown in Figure 3.3, is to foresee the following word in a grouping, given the former words.

One of the major issues in sequence modeling is to handle long-term dependency. For instance in the sentence given here, we want to predict the word in the blank space. By using the previous words with window size of two, three, or four, we can't predict the correct word. Only by using the information from the previous word *France* and future word *language*, we can guess accurately.

"I had a wonderful time in France and learned some of the _____ language."
To model sequences, we need to do the following:

mice/NNS eat/VBP cheese/NN

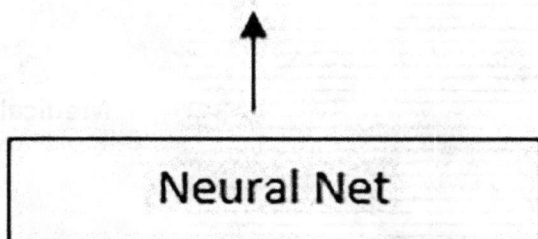

↑

Neural Net

↑

Mice eat Cheese

FIGURE 3.2 Sequence Labeling.

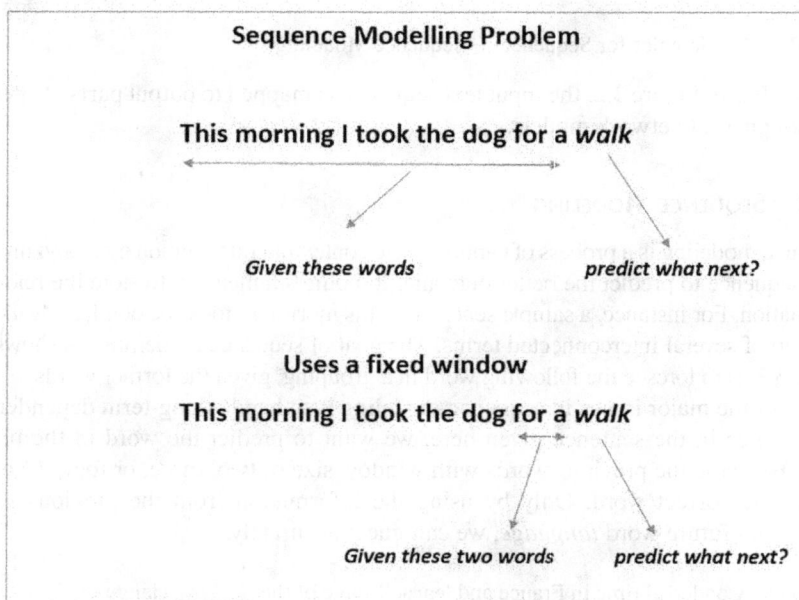

Sequence Modelling Problem

This morning I took the dog for a *walk*

Given these words   predict what next?

Uses a fixed window

This morning I took the dog for a *walk*

Given these two words   predict what next?

FIGURE 3.3 Next Word Prediction.

Bag of Words don't preserve order

The food was good, not bad at all

Versus

The food was bad, not good at all

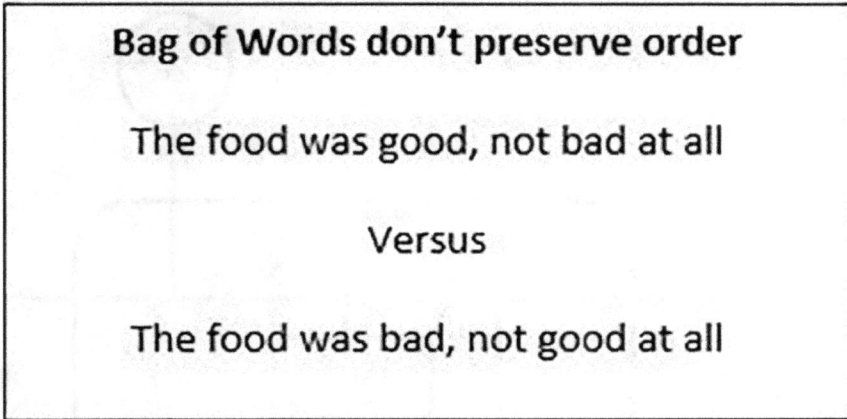

FIGURE 3.4 Sequence Ordering.

- To deal with variable length sequence
- To maintain sequence order, for instance, bag of words don't preserve order as shown with a simple example in Figure 3.4.
- To keep track of long-term dependencies
- To share parameters across the sequence

3.3 RECURRENT NEURAL NETWORKS

RNN helps to process sequential data or time series data. A simple RNN takes input from sequential data $\{x_1, x_2, \ldots x_n\}$ where n is the time step and produces a sequence $\{y_1, y_2, \ldots, y_n\}$. As the name indicates, RNN is made up of recurrent neurons with two inputs at each time step such as current input and previous output as shown in Figure 3.5. RNN consists of a feedback loop which is attached to neurons' previous outcome. RNN applies a recurrent relation at every time step to process a sequence.

$$h_t = f_w\left(h_{t-1}, x_t\right) \tag{1}$$

$$h_t = \tanh\left(Uh_{t-1}, Wx_t\right) \tag{2}$$

W represents the weight matrix, U represents the transition matrix, and h_t represents the hidden state at time step t and h_{t-1} represents the hidden state at time step t−1. The nonlinearity introduced by the activation function tanh overcomes the vanishing gradient problem.

To store the information from previous time step, it has an in-built memory.

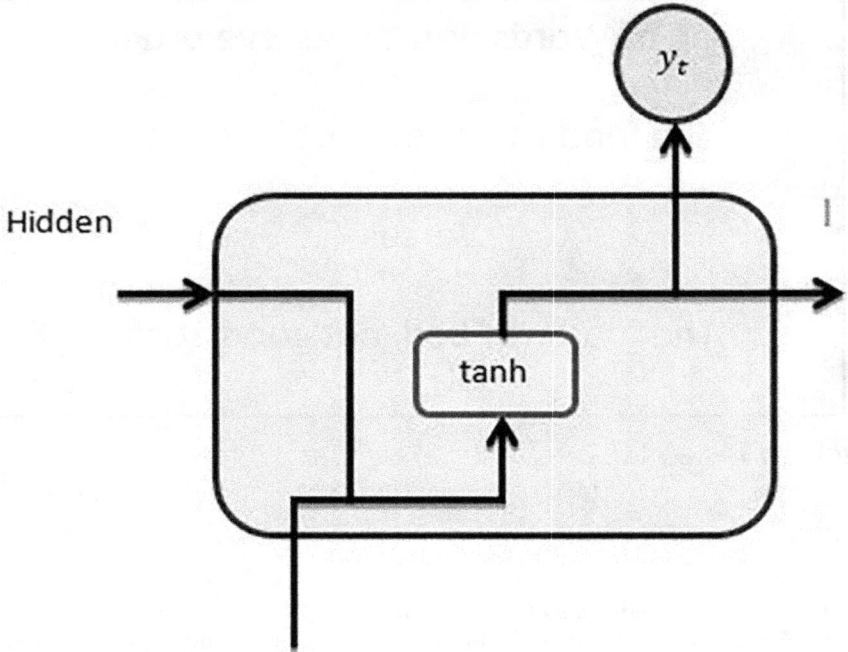

FIGURE 3.5 Single RNN Unit.

3.3.1 Unrolling RNN

A graphical representation of an unrolling RNN is shown in Figure 3.6. Initially, at time step 0, x_0 is given as input to RNN cell to product output y_0. As propagation, the next RNN cell receives the hidden state from time step 0. RNN works in such a way that the output of the current time step is dependent on both the input and the results of the previous time step. The parameters U and V are shared across the RNN layer.

The different topologies of RNN are shown in Figure 3.7.

- One to one: It is similar to CNN which accepts the input of fixed size vector and converts into one discrete output class label. Example being pattern recognition.
- One to many: One input of fixed size vector is converted to a sequence of output. Example being image captioning.
- Many to one: Sentiment analysis is an example for many to one which takes sequential text as input and identifies the output.
- Many to many: Machine translation is a good example for many to many mapping.

The two major variants of RNN are gated recurrent unit (GRU) and LSTM.

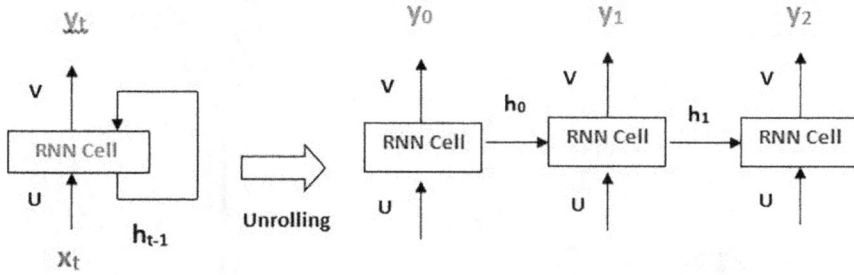

FIGURE 3.6   Unrolling RNN.

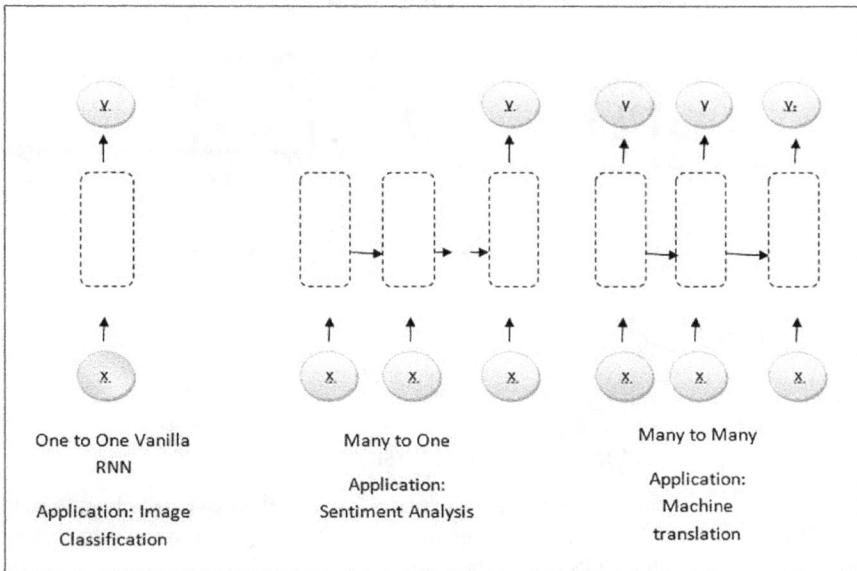

FIGURE 3.7   Variants of RNN Architecture.

- GRU: It is composed of two gates which control the information to be retained.
  - Reset gate
  - Update gate
- LSTM: A single LSTM unit is depicted in Figure 3.8. LSTM replaces the single tanh gate with four gates as follows.
  - Forget gate: Forgets the irrelevant parts of the previous state.
  - Save: The LSTM saves relevant data in the cell state.
  - Update: Only changes cell state values when necessary.
  - Output: It determines which data must be transferred to the next state.

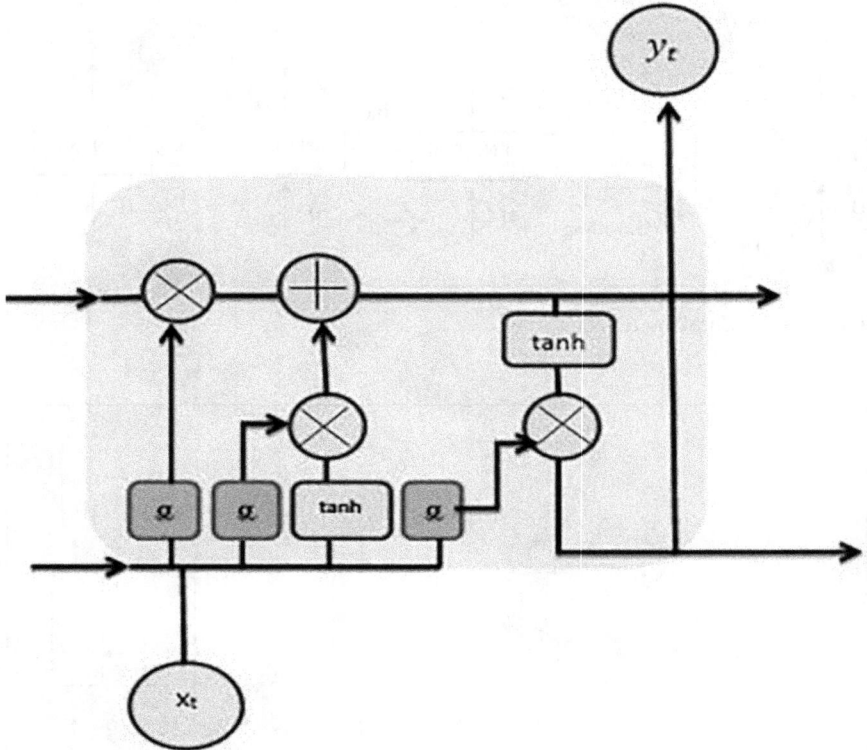

FIGURE 3.8   LSTM Unit.

3.3.2   RNN-BASED POS TAGGING USE CASE

Parts of speech tagging is a text pre-processing mechanism extensively integrated with NLP applications. The POS tagger receives a language sequence as input and generates the associated POS tags. It is a very good example of many to many RNN topologies. The working of RNN for an NLP use case, parts of speech tagging (POS) is given in Figure 3.9.

3.3.3   CHALLENGES IN RNN

RNN fails to handle varied length sequence—for instance, in machine translation if we want to translate English to Tamil as shown in Figure 3.10. The length of input sequence is three whereas that of the output sequence is two. RNN is not suitable for such varied length sequences.

3.4   ATTENTION MECHANISMS

The ability to focus on one thing while ignoring others is critical for intellect direction. This allows us to recall only one experience rather than all of them to pursue

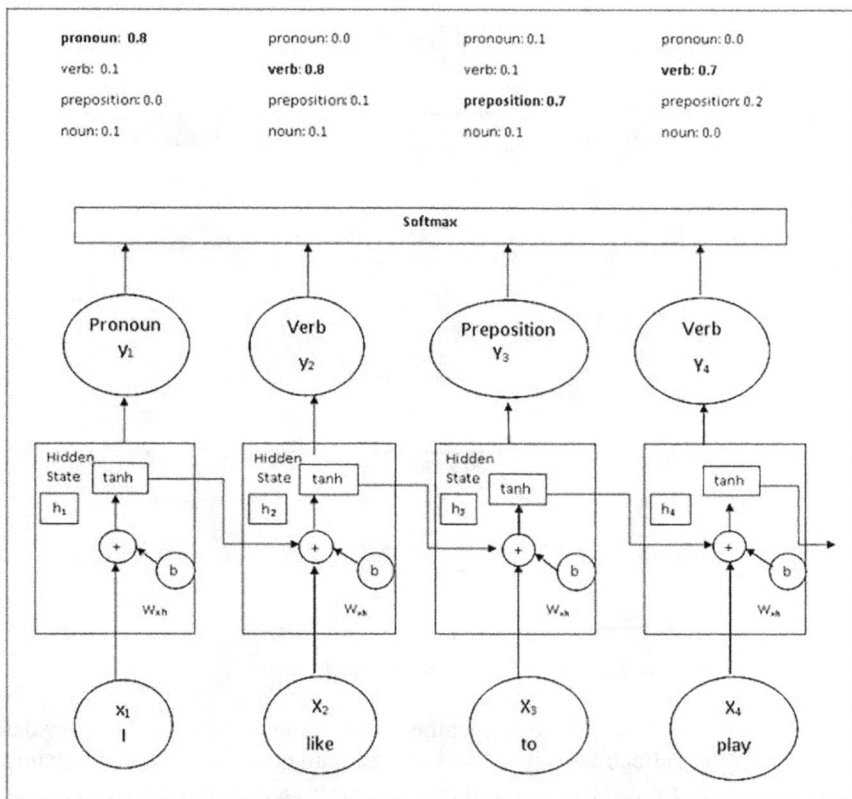

FIGURE 3.9   POS Tagging Using RNN.

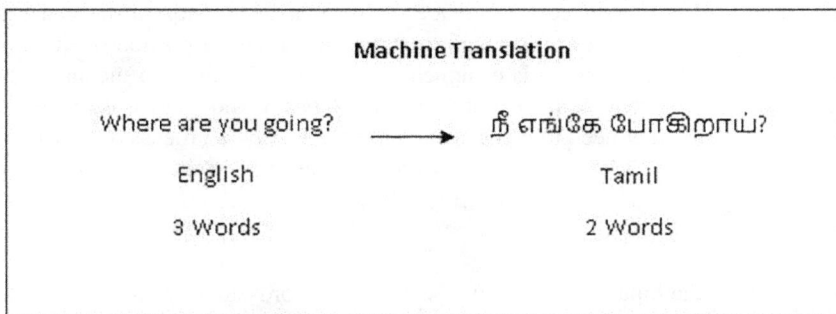

FIGURE 3.10   Machine Translation.

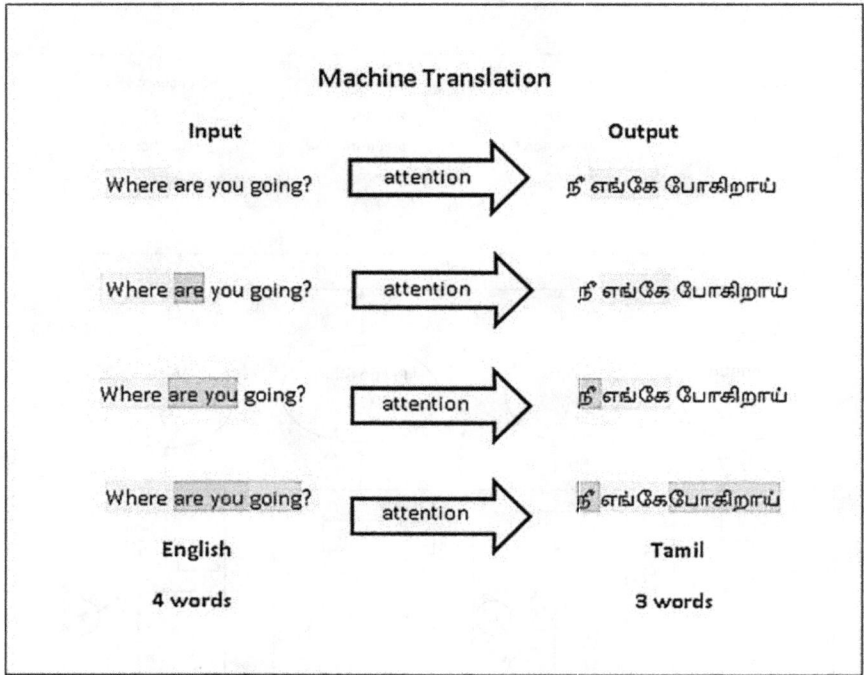

FIGURE 3.11   Machine Translation with Attention Mechanism.

only one thought at a time and to extract the important information from noisy data, giving more strong attention to some part of data than others. Attention mechanism is widely accepted for machine translation applications as given in Figure 3.11.

3.4.1   SELF-ATTENTION MECHANISM

Self-attention mechanism is inspired by the working of the human brain in such a way that when humans see a sequence of input or an image, the human brain effortlessly gives attention to a particular or the most relevant portion of the input. The same way attention model is designed. The name self-attention means it gives attention to itself. Given a sequence of words, it allows the model to look at what are all the other words that need to be paid attention to while you process one particular word. Figures 3.12, 3.13, and 3.14 depict the working of self-attention mechanism. The following are the steps in performing self-attention mechanism.
Steps:

1. Tokenize the input words and $[w_1, w_2, \ldots . w_n]$, word-embedding vectors are generated.
2. The input vector is multiplied with weight matrixes such as w_q, w_k, w_v to produce query $[q_1, q_2, \ldots q_n]$, key $[k_1, k_2, \ldots k_n]$, and value $[v_1, v_2, \ldots v_n]$ vectors.

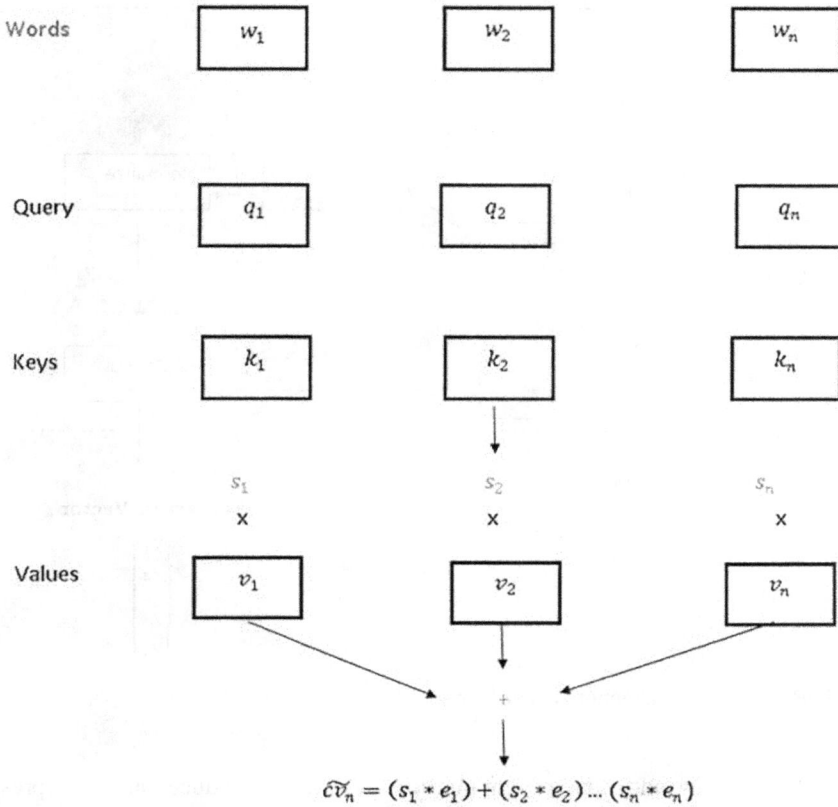

FIGURE 3.12   Contextually Analyzing the Words Using Self-attention.

3. The inner product of query and key vector produces attention score or rank which is $[s_1, s_2, \ldots s_n]$. If the score is higher, it implies similar words or else dissimilar words.
4. The score values are then multiplied with value vector and added to produce the context vector $\tilde{c}v_n$.

$$\tilde{c}v_n = \left(s_1 * e_1\right) + \left(s_2 * e_2\right) \ldots \left(s_n * e_n\right)$$

3.4.2   MULTI-HEAD ATTENTION MECHANISM

Multi-head attention is basically for loop over the self-attention mechanism as given in Figure 3.15. The calculation of self-attention for a sequence each time is called a head. Thus, the term "multi-head attention" refers to the process of performing self-attention multiple times at the same time. After that, the independent attention outputs are combined and linearly converted into the desired dimension. Multiple

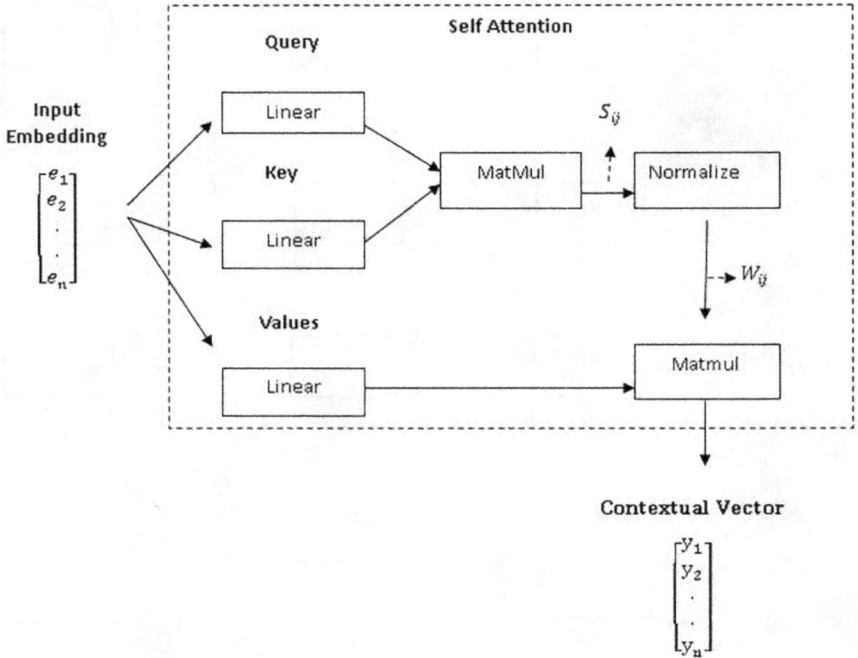

FIGURE 3.13   Self-attention Mechanism.

sets of query, key, and value weight matrices are used to produce multiple representations of query, keys and value vectors vectors as shown in figure 3.16.

$$MultiHead\ (Q,\ K,\ V)\ [head_1..., head_n]\ W_0$$

$$where\ head_i = Attention\left(QW_i^Q, KW_i^K, VW_i^V\right)$$

3.4.3   BAHDANAU ATTENTION

Bahdanau architecture uses encoder decoder architecture using bidirectional recurrent neural network (Bi-RNN) which peruses input sentence in forward heading to
deliver forward secret state $\overrightarrow{h_i}$ and then move in the backward direction to produce
backward hidden state. $\overleftarrow{h_i}$.

$$h_i = \left[\overrightarrow{h_i^T} : \overleftarrow{h_i^T}\right]^T$$

The context vector is the output of the encoder and is fed into the decoder architecture. It makes use of additive attention as shown in Figure 3.17 to produce the
context vector.

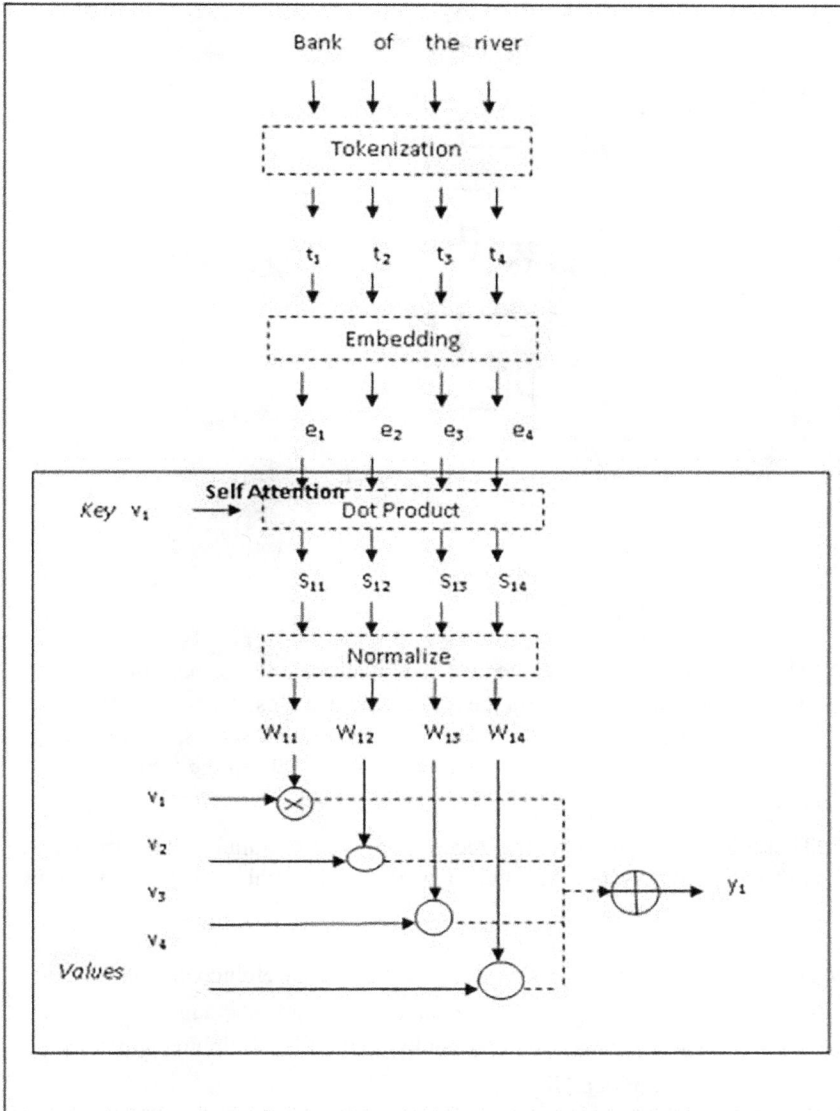

FIGURE 3.14 Attention Computation with Key, Query, and Value Vectors.

3.4.4 LUONG ATTENTION

1. The encoder produces the hidden states, $H = h_i, i = 1,...,T$, from the information sentence.

2. The present decoder hidden state is derived as: $s_t = RNN_{decoder}(s_{t-1}, y_{t-1})$. Here, s_{t-1} implies the previous hidden decoder state, and y_{t-1} the previous decoder output.

Multi-Head Attention

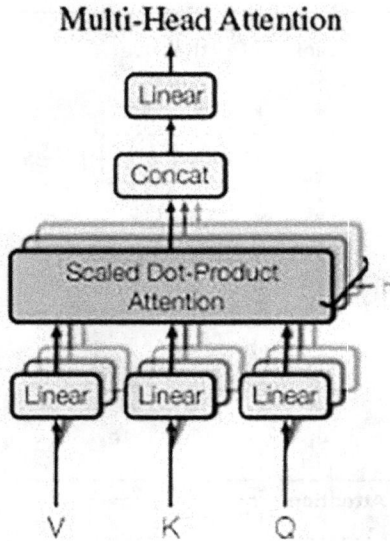

FIGURE 3.15   Multi-head Attention.

Source: A. Vaswami et al.

3. Alignment model, a(.), utilizes the explanations and the ongoing decoder-stowed-away state to process the arrangement scores: $e_{t,i} = a\left(s_t, h_i\right)$.

4. A softmax capability is applied to the arrangement scores, really normalizing them into weight values in a reach somewhere in the range of 0 to 1: $\alpha_{t,i} = softmax\left(e_t, i\right)$.

5. These loads, along with the recently figured comments, are utilized to create a setting vector through a weighted amount of the explanations:
$$c_t = \sum_{i=1}^{T} \alpha_{t,i} h_i$$

6. An attentional secret state is registered in view of a weighted link of the setting vector and the ongoing decoder-stowed-away state: $\tilde{s}_t = \tanh\left(W_c\left[c_t; s_t\right]\right)$

7. The decoder produces a final output by feeding it a weighted attention hidden state: $y_t = softmax\left(W_y\, \tilde{s}_t\right)$.

8. Stages 2–7 are rehashed for the rest of the arrangement.

3.4.5   GLOBAL ATTENTION VERSUS LOCAL ATTENTION

Global Attention: Global attention, as the name suggest for each and every input tokens $\{i_1, i_2, \ldots i_n\}$, is calculated for every other tokens. BERT architecture is a well-known example for global attention. Obviously, for lengthy sequence, the global attention architecture is computationally expensive. The working principle behind global architecture is given in Figure 3.18.

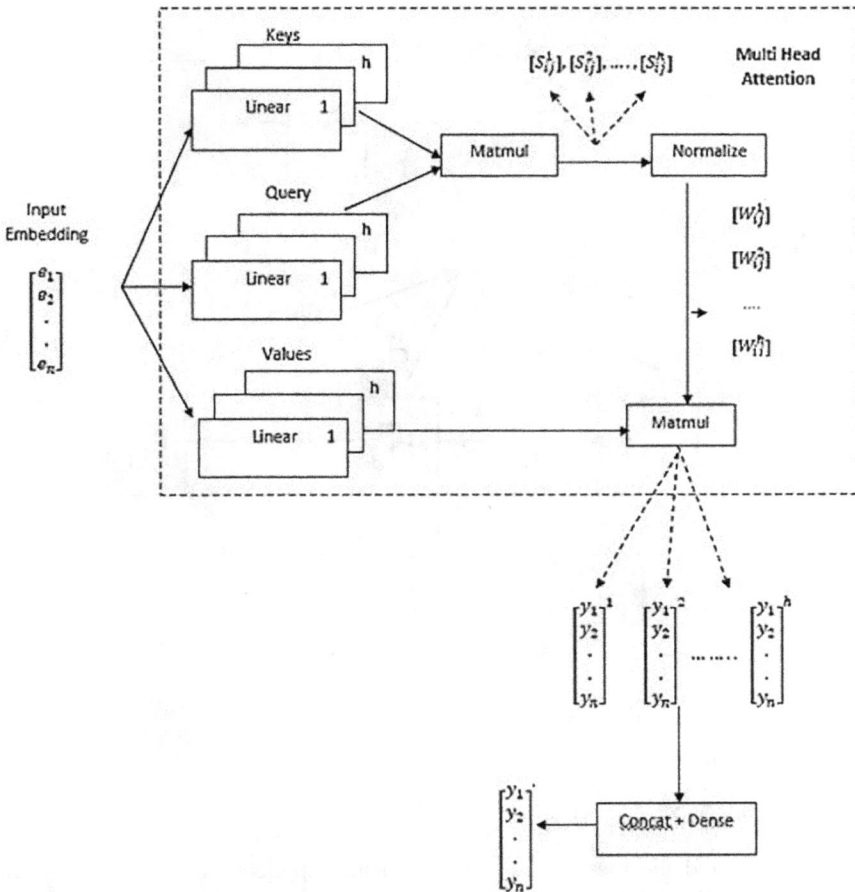

FIGURE 3.16 Multi-head Attention with K, Q, and V.

Local Attention: To address the drawback of global attention and to minimize the computational cost, local attention mechanism comes into picture. Instead of paying attention to all the input tokens, only subset of tokens is considered, which is definitely a computationally efficient process. There are three variants of local attentions such as

- Local attention with window size 1
- Local attention with window size 2
- Local attention with random samples

Figures 3.19 and 3.20 show the local attention working with widow size 1 and 2, respectively. Figure 3.21 shows the working behind random attention.

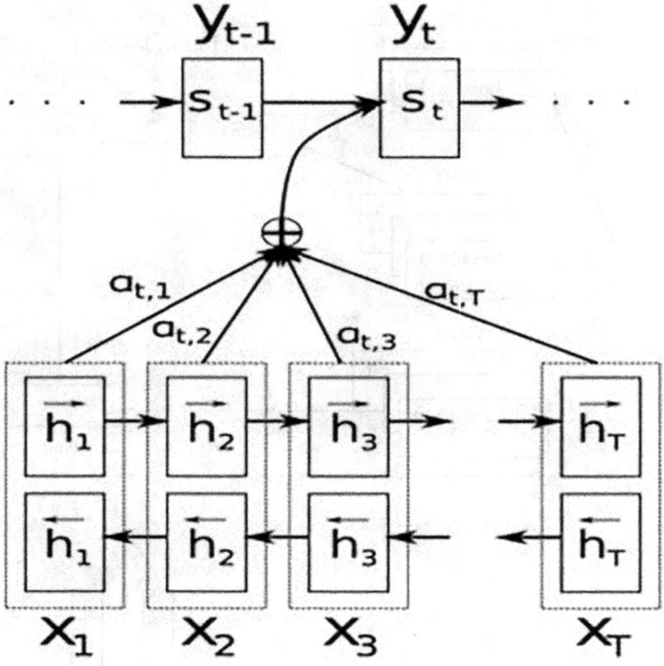

FIGURE 3.17   Bahdanau Attention.

Source: Dzmitry Bahdanau et al.

3.4.6   HIERARCHICAL ATTENTION

Let us understand the hierarchical attention using applications like chat box and document classification. Nowadays, the most widely used NLP-based application is chat box. Figure 3.22 shows typical chat box dialogs. The dialog is a sequence of a sequence which in turn consists of utterances between the user and the bot. Each utterance is in turn a sequence of words. Hierarchical attention (HAN) is a well-suited model for sequence of sequence problems. HAN consists of two-level attention layers as shown in the architecture in the Figure 3.23. First, we need to attend to the most important and informative words in a sentence:

The components of HAN are

- Word Encoder
- Word Attention
- Sentence Encoder
- Sentence Attention

3.5   TRANSFORMER MODEL

Transformers are a fairly new family of neural network architectures. BERT is a particular huge transformer-veiled-language model. A language model is a factual model of the likelihood of a sentence or expression. To understand what language

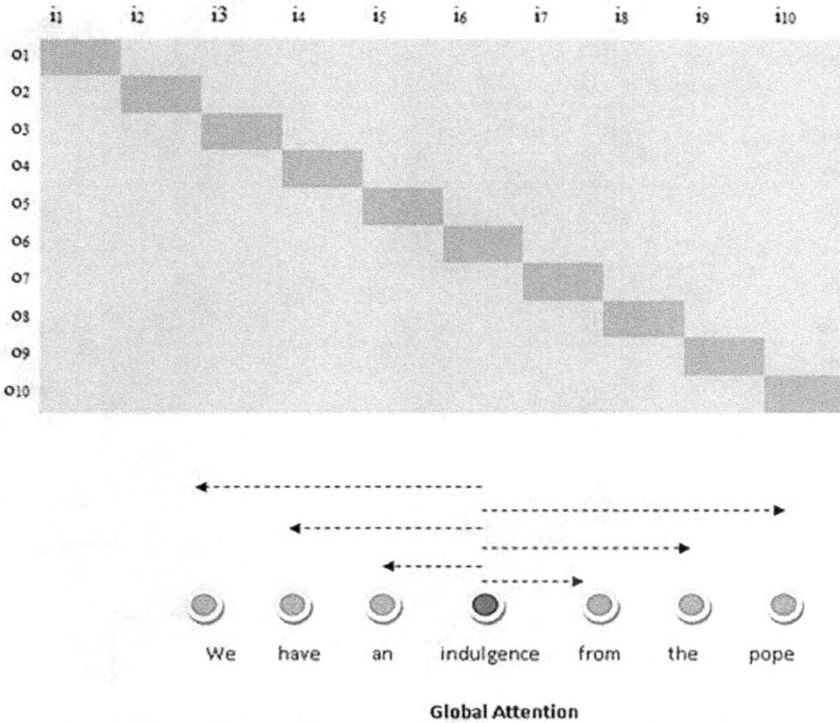

FIGURE 3.18 Global Attention.

model is, let us consider an example as shown in Figure 3.24. The probability of the phrase "Tensorflow is open source" is greater than the probability of phrase "Source Tensorflow is open," given some training corpus.

A masked language model is a smidgen unique, so rather than distinguishing the likelihood of an entire expression, the most common way of preparing the model is by having a fill in the clear. Concealed language models are helpful on the grounds that there is one approach to doing relevant word embedding. BERT is extremely large: the larger version has 340 million trainable parameters when compared to word-embedding model ELMo, which had only 93 million.

The original transformer architecture is shown in Figure 3.25. The model utilizes conventional succession to grouping where you have an encoder, which takes information and transforms it into embedding, and a decoder that takes those embeddings and transforms them into string yield.

3.5.1 BIDIRECTIONAL ENCODER, REPRESENTATIONS, AND TRANSFORMERS (BERT)

BERT stands for Bidirectional Encoder, Representations, and Transformers. BERT model is "deeply bidirectional." Bidirectional alludes to the way that BERT accumulates information from both the left and right sides of a symbolic setting during the preparation stage. To fully comprehend the meaning of a language, a model must be

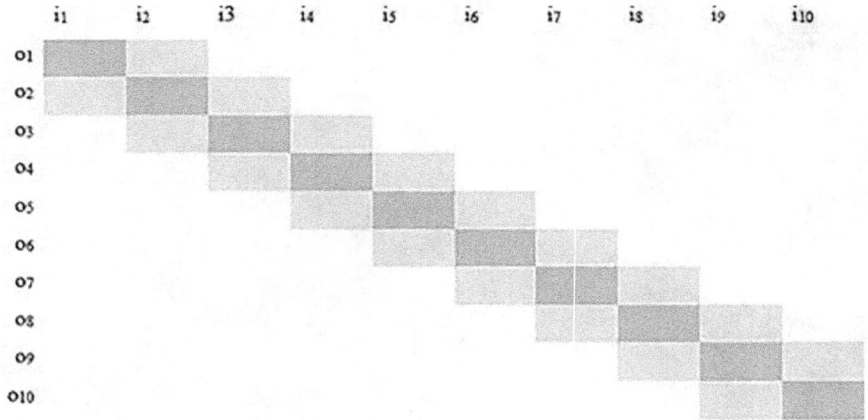

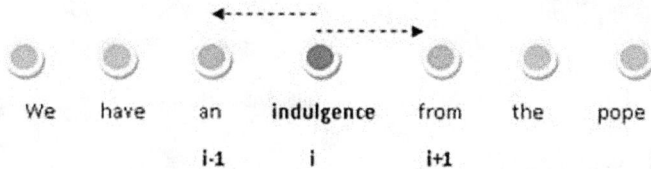

FIGURE 3.19 Local Attention with Window Size 1.

bidirectional. Take a look at Figure 3.26 for an example of this. The term "bank" appears in both statements in this example.

- **Bidirectional:** This model reads text from both directions (left and right) to gain a better understanding of text.
- **Encoder:** Encoder Decoder model is used in NLP where the task is feeding to the input of encoder and the output is taken from the decoder.
- **Representation:** Encoder Decoder architecture is represented using transformers.
- **Transformers:** A key part of the transformers is the multi-head attention block. Transformers is a mix of attention, standardization, and veiled consideration in the decoder stage.

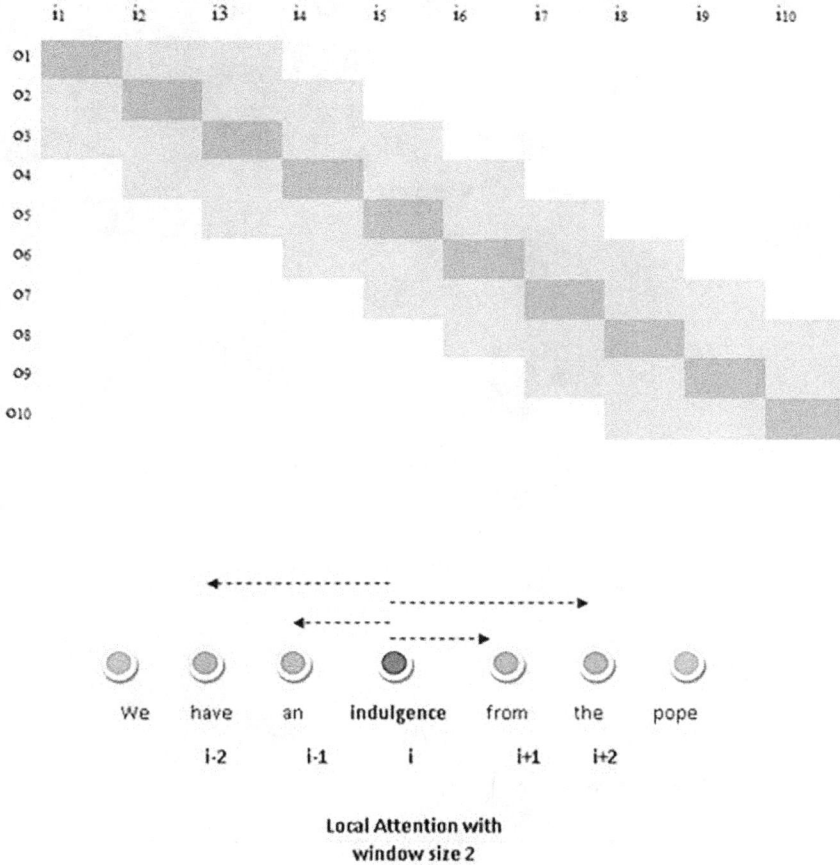

FIGURE 3.20 Local Attention with Window Size 2.

BERT engineering is a tad different, which takes various encoders and stacks them on top of one another. BERT base and BERT huge are the two variations of BERT as displayed in Figure 3.27. The CLS token in front is utilized to address the order of a particular piece of information, though SEP tokens are utilized toward the finish of each and every information grouping.

The BERT design is to utilize an extra classifier to adjust the first model, where we really update the loads in the first BERT model as given in Figure 3.28. There are several architectures proposed by extending the BERT model such as:

- RoBERTa
- DistilBERT
- AlBERT

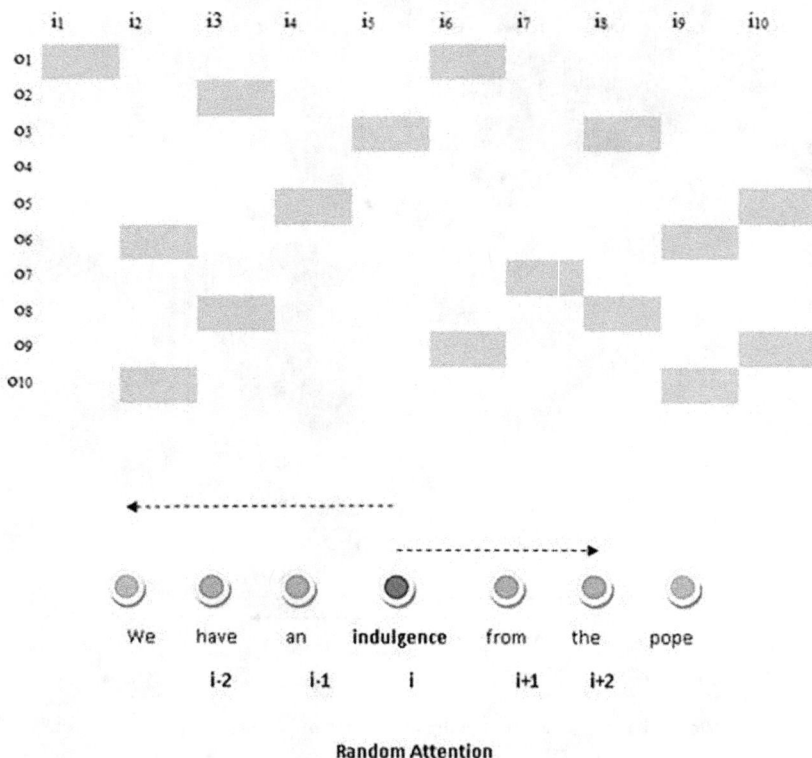

FIGURE 3.21   Random Attention.

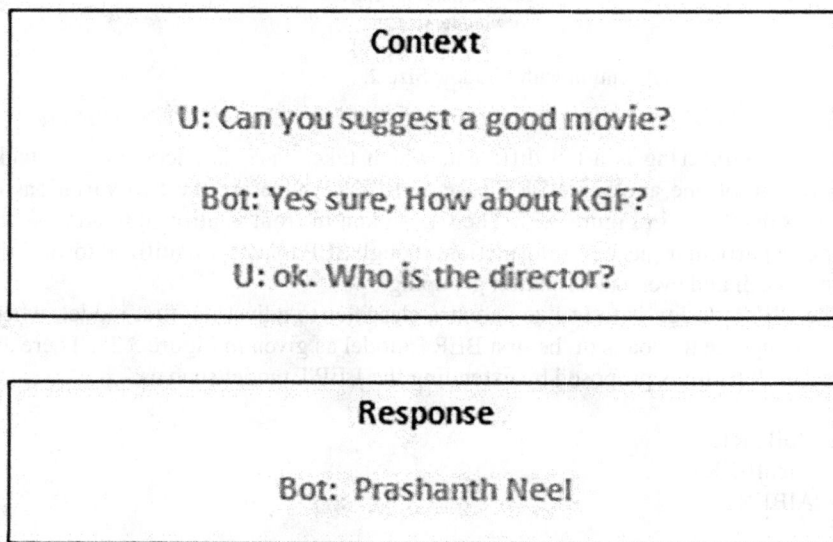

Context

U: Can you suggest a good movie?

Bot: Yes sure, How about KGF?

U: ok. Who is the director?

Response

Bot: Prashanth Neel

FIGURE 3.22   Chatbot.

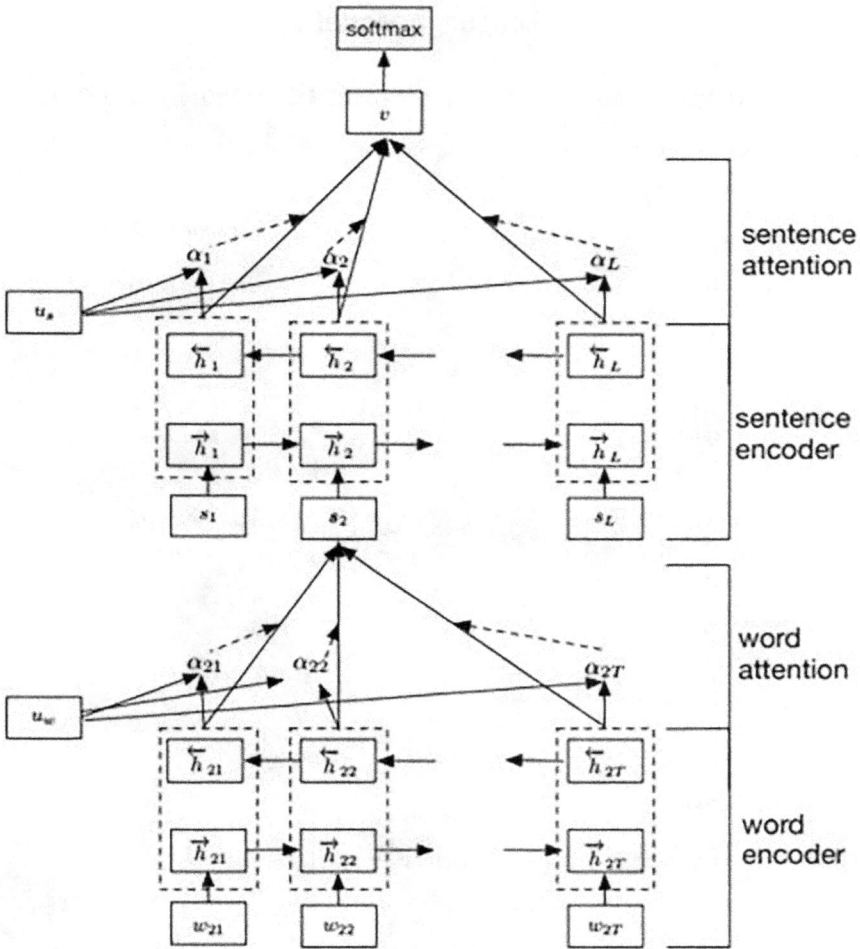

FIGURE 3.23 Hierarchical Attention.

Source: Zichao Yang et al.

Architectures in other languages are as follows:

- CamemBERT (French)
- AraBERT (Arabic)
- Mbert (multilingual)

Some of the other BERT variants are given as follows:

- patentBERT, a BERT model designed specifically for patent classification.
- docBERT, a BERT model optimized for document classification.
- bioBERT, a biomedical text mining model pretrained to represent biomedical language.

Language Model

P (Tensorflow is open source) > P (Source Tensorflow is open)

FIGURE 3.24   Language Model.

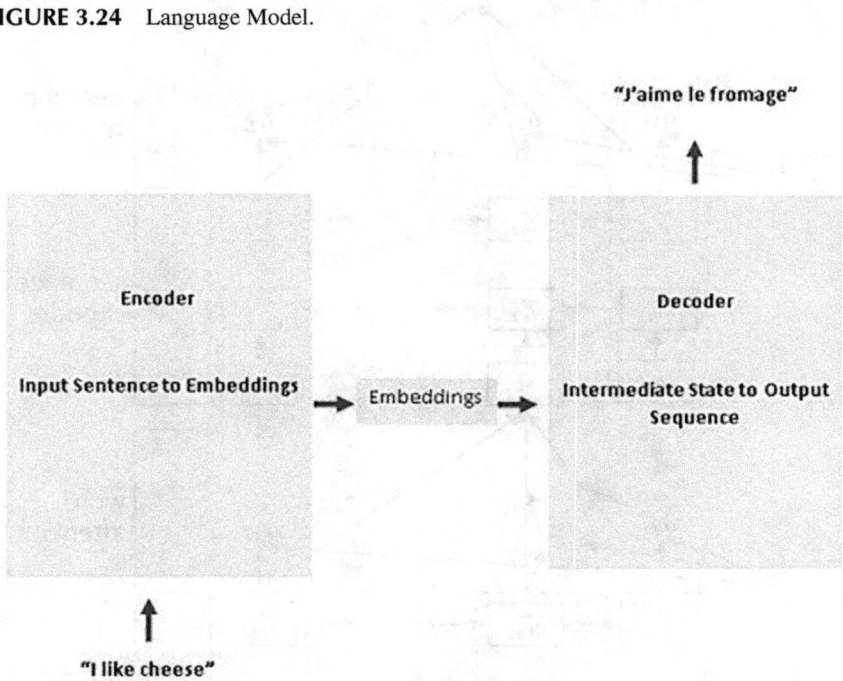

FIGURE 3.25   Encoder Decoder-based Machine Translation.

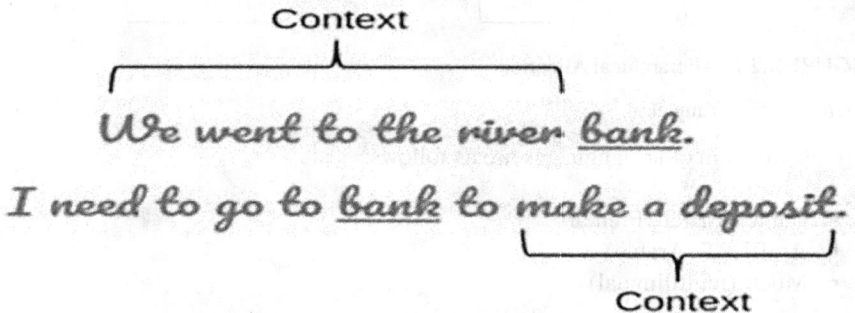

FIGURE 3.26   Contextual Learning by BERT.

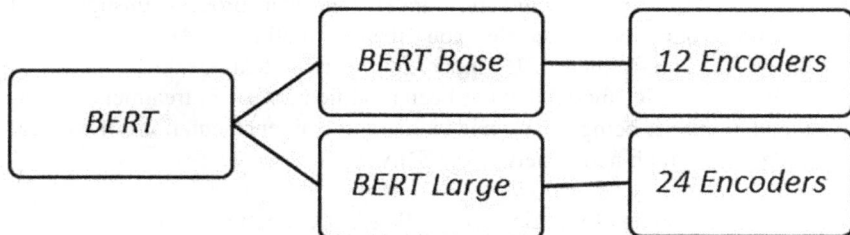

FIGURE 3.27 Variants of BERT.

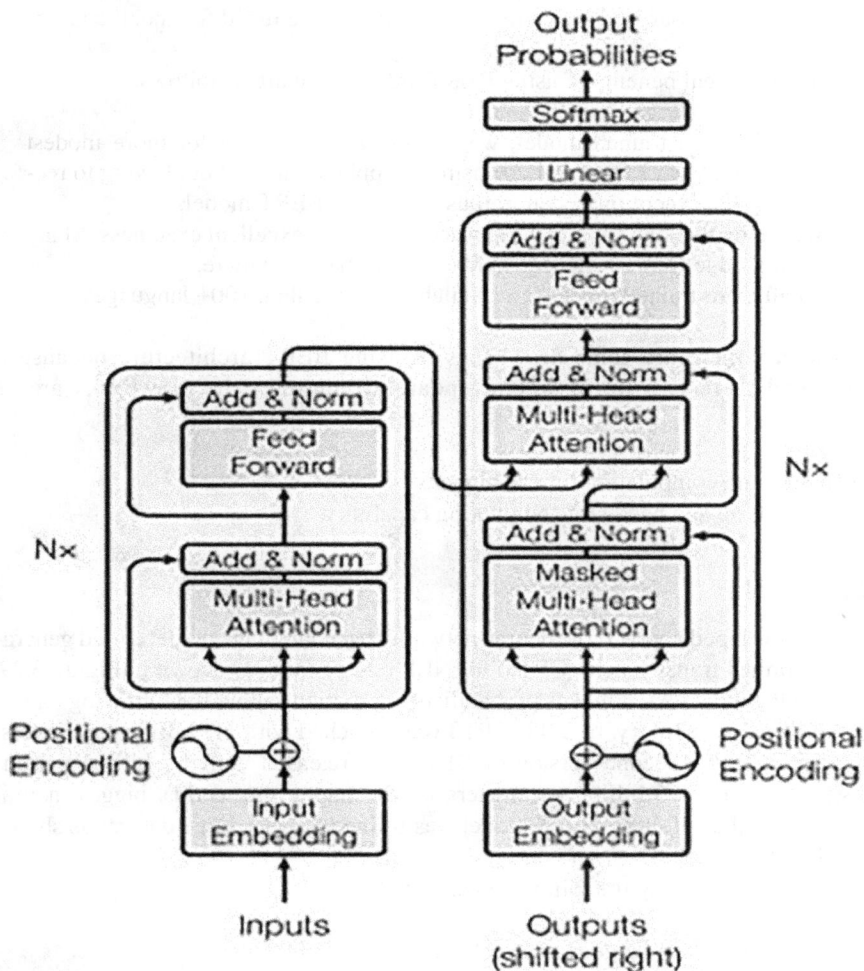

FIGURE 3.28 BERT Architecture.

Source: A. Vaswami et al.

- VideoBERT, a visual-linguistic model that can process unsupervised learning from a lot of unlabeled YouTube information.
- SciBERT, a pre-trained BERT model for scientific text.
- G-BERT, a BERT model that has been modified for use in treatment recommendation after being pre-trained on hierarchical representations of medical codes with graph neural networks (GNNs).

The following are some of the use cases of BERT architecture:

- To convert text input into numeric representation (i.e., to replace word embeddings).
- As a general-purpose pre-trained model that is fine tuned for specific tasks.

Some of the real benefits of using BERT architecture are as follows:

- As it is a pre trained model, we can utilize this model for more modest assignments of explicit and downstream applications without having to recompute the exceptionally enormous and costly BERT model.
- With effective fine-tuning, we can accomplish an excellent exactness. Many cutting-edge frameworks consolidate BERT here and there.
- Finally, pre-trained models are available in more than 100+ languages.

However, there are some drawbacks of using BERT architecture because it is big, and there is a lot of weight updating required. The drawbacks are as follows:

- Huge and computation time is higher.
- Expensive and need large computing capability.

3.5.2   GPT3

OpenAI developed a very large neural-network-based language model called generative pre-trained transformers (GPT3) for NLP-related tasks as shown in Figure 3.29. Based on the input text, it makes probabilistic predictions about the following tokens from a known vocabulary. In 2018, GPT3 was launched with 117 billion parameters. Later, in 2019, the second version of GPT3 was released with 1.5 billion parameters. GPT3 with 175 billion parameters is one among the world's biggest neural network model as of 2021. GPT3 model was trained on very large dataset as shown in Table 3.1.

It is capable of doing the following tasks such as,

- Creating human language natural text.
- Creating realistic human text.
- Used in automated conversational tasks such as chatbot.

Some of the benefits of GPT3 language models are as follows:

Input: *Recite the Newton's third law*

↓

Generative Pre Trained Transformers (GPT 3)

↓

Output: *To every action there is always opposed an equal reaction; or, the mutual actions of two bodies upon each other are always equal, and directed to contrary parts*

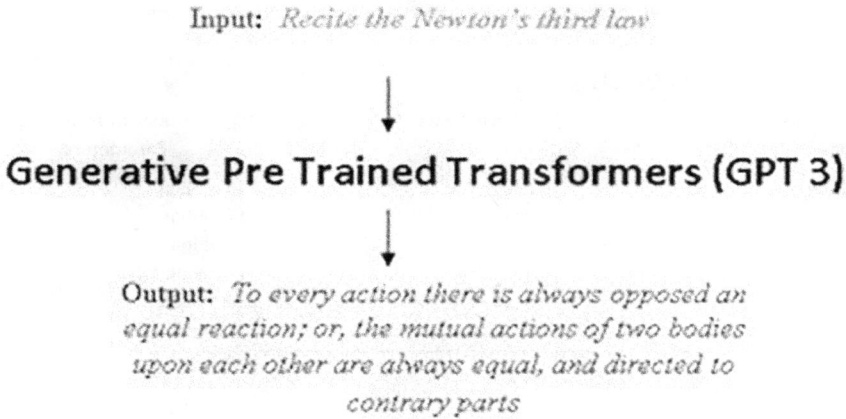

FIGURE 3.29 GPT3 Language Model.

TABLE 3.1
Dataset Used in GPT3 Training

Dataset	# Tokens	Content
Common Crawl	410 billion	8+ years of raw web page data, metadata extracts, and text extracts with light filtering
WebText2	19 billion	All incoming Reddit links from posts with three or more upvotes
Books1	12 billion	
Books2	55 billion	
Wikipedia	3 billion	English-language Wikipedia pages

- They have more fluent-sounding text.
- They are single pre-trained models that are extremely adaptable and can be used for a variety of tasks.
- They are conceptually simple to use.

The drawbacks of GPT3 are

- Really only good for English
- Big and expensive
- Closed API access
- Difficult to compare to previous models
- Unpredictable output

Table 3.2 explains the widely used language models with their number of trainable parameters.

TABLE 3.2

List of Language Models

Language Models	Released By	Number of Trainable Parameters
Bert Large	Google	340 million
GPT	Open AI	117 million
GPT2	Open AI	1.5 billion
GPT3	Open AI	175 billion
T5	Google	220 million
Turing NLP	Microsoft	17 billion

3.6 SUMMARY

This chapter explores the various models for sequential data analysis such as RNN and attention-based models with examples in detail. The variants of RNN are discussed and introduce language modeling and transformer-based models used in NLP applications. Models like BERT and GPT3 were also discussed in detail. Next chapter focuses on various applications of NLP that can be implemented using the models and techniques described in this chapter.

BIBLIOGRAPHY

Bahdanau, D., Cho, K. H., & Bengio, Y. (2015). *Neural machine translation by jointly learning to align and translate*. Paper presented at 3rd International Conference on Learning Representations. San Diego, CA: ICLR.

Bengio, Y., Goodfellow, I., & Courville, A. (2017). *Deep learning*, vol. 1. Cambridge, MA: The MIT press.

Chung, J., Kastner, K., Dinh, L., Goel, K., Courville, A. C., & Bengio, Y. (2015). *Advances in neural information processing systems*, vol. 28. New York: Curran Associates, Inc (A Recurrent Latent Variable Model for Sequential Data).

Devlin, J., Chang, M., Lee, K., & Toutanova, K. (2019). *Proceedings of the 2019 conference of the North American chapter of the association for computational linguistics: Human language technologies*, vol. 1 (Long and Short Papers, pp. 4171–4186). Minneapolis: Association for Computational Linguistics.

Dou, Y., Forbes, M., Koncel-Kedziorski, R., Smith, N., & Choi, Y. (2022). Is GPT-3 text indistinguishable from human text? Scarecrow: A framework for scrutinizing machine text. *Proceedings of the 60th annual meeting of the association for computational linguistics*, vol. 1 (Long Papers, pp. 7250–7274). Dublin: Association for Computational Linguistics.

Floridi, L., & Chiriatti, M. (2020). GPT-3: Its nature, scope, limits, and consequences. *Minds & Machines* 30, 681–694.

Graves, A. (2012). Supervised sequence labelling. In: *Supervised sequence labelling with recurrent neural networks* (Studies in Computational Intelligence), vol. 385. Berlin and Heidelberg: Springer.

Graves, A., Mohamed, A. R., & Hinton, G. (2013). Speech recognition with deep recurrent neural networks. In: *IEEE international conference on acoustics. Speech and signal processing (ICASSP)* (pp. 6645–6649). Vancouver, Canada: IEEE.

Han, Z., Ian, G., Dimitris, M., & Augustu⁵, O. (2019). *Proceedings of the 36th international conference on machine learning* (pp. 7354–7363). New York: PMLR.

Sneha, C., Mithal, V., Polatkan, G., & Ramanath, R. (2021). An attentive survey of attention models. *ACM Transactions on Intelligent Systems and Technology (TIST)* 12(5), 1–32.

TensorFlow. Recurrent neural network s. *TensorFlow.* www.tensorflow.org/tutorials/recurrent.

Vaswani, A., et al. (2017). *Attention is all you need, advances in neural information processing systems.* Kolkata: NIPS.

Zichao, Y., Yang, D., Dyer, C., He, X., Smola, A., & Hovy, E. (2016). Hierarchical attention networks for document classification. In: *Proceedings of the 2016 conference of the North American chapter of the association for computational linguistics: Human language technologies* (pp. 1480–1489). San Diego, CA: Association for Computational Linguistics.

4 Applications of Natural Language Processing

LEARNING OUTCOMES

After reading this chapter, you will be able to:

- Understand the basic building blocks of NLP applications.
- Identify the pipeline of applications like sentiment analysis, text classification, and question-answering System.
- Identify the appropriate steps to be chosen for the NLP use cases.

4.1 INTRODUCTION

The applications of Natural Language Processing are extensively used in many fields in day-to-day life. The search engines use NLP to actively understand the user requests and provide the solutions in a faster manner. The smart assistants like Alexa and Siri rely on NLP. NLP is used in business-related fields by providing the facts and figures of each and every year growth and production. It evaluates text sources from emails to social media and many more. Unstructured text and communication are transformed into usable data for analysis utilizing a variety of linguistic, statistical, and machine learning techniques by NLP text analytics. NLP and AI tools can automatically comprehend, interpret, and classify unstructured text using text classification. Data is arranged on the basis of corresponding tags and categories using the Natural Language Processing algorithms. The text extraction uses NLP, often referred to as named entity recognition, and may automatically detect particular named entities within text. Text summarization uses NLP to efficiently process the input and obtain very important information. It is widely used in educational sector, research, or healthcare environments. NLP can provide a paraphrased summary of a text by focusing on specific key phrases within the text or by determining meanings and conclusions. Market intelligence uses NLP for separating subjects, sentiment, keywords, and intent in unstructured data from any type of text or consumer communication. The feature of intent classification enables businesses to more precisely determine the text's intention through their emails, social media posts, and other communication, and it can help customer care teams and sales teams.

Financial firms can use sentiment analysis to examine more market research and data and then use the knowledge gained to streamline risk management and make better investment decisions. Banks and other security organizations can use NLP to spot instances of money laundering or other frauds. It is possible to use natural language processing to assist insurance companies in spotting fraudulent claims. AI can spot signs of frauds and flag such claims for additional examination by examining customer communications and even social media profiles.

DOI: 10.1201/9781003348689-4

Insurance companies use natural language processing to monitor the highly competitive insurance market environment. It can better understand what their rivals are doing by utilizing text mining and market intelligence tools, and they can plan what products to launch to keep up with or outpace their rivals. NLP analyses the shipment manufactures and provides information about the shortage of supply chain to improve the automation and also manufacturing pipeline. With the help of this information, they can improve specific steps in the procedure or adjust the logistics to increase the efficiency. Sentiment analysis should be used, especially by retailers. Retail businesses can improve the success of each of their company activities, from product release and marketing, by measuring customer sentiments regarding their brands or items. NLP makes use of social media comments, customer reviews, and other sources to transform this information into useful information that merchants can utilize to develop their brand and address their problems. The potential uses of natural language processing in the healthcare industry are enormous, and they are only just beginning. It is now assisting scientists working to combat the COVID-19 pandemic in a number of ways, including by examining incoming emails and live chat data from patient help lines to identify those who may be exhibiting COVID-19 symptoms. This has made it possible for doctors to proactively prioritize patients and expedite patient admission to hospitals.

This chapter discusses the different types of applications involving Natural Language Processing that is extensively used, and they are word sense disambiguation (WSD), word sense induction, text classification, sentiment analysis, spam email classification, question answering, information retrieval, entity linking, chatbots, and dialog system.

4.2   WORD SENSE DISAMBIGUATION

The WSD is one of the basic applications in Natural Language Processing. The WSD exhibits the capacity to ascertain which meaning of a word is triggered by use in a specific context. The WSD is a disambiguation between different senses of words. The WSD is performed on basis of the task, domain, and size of the word and sense sets. The baseline of the WSD is that, given a new word, assign the most frequent sense to it on the basis of counts from a training corpus. For example, in order to perform automated medical articles, disambiguating things on the basis of the medical subject headings need to be performed. In order to disambiguate a small number of words, the small traditional supervised classification techniques work well. Similarly, for disambiguating huge sets of words, a large and complex number of models need to be built. The popular sense-tagged corpora such as SemCor and Senseval Corpora are used to train the models. The prior knowledge of WordSense and WordNet is required to study about the WSD. In the following sections, the concepts of WordSense and WordNet is elaborated.

4.2.1   WORD SENSES

A sense, sometimes known as a word sense, is a distinct representation of one part of a word's meaning. When working on tasks that include meaning, understanding the relationship between two senses might be crucial. For example, think about the antonymy relationship. When two words, such as long and short or up and down, have the opposing meanings, they are said to be anagrams. It is crucial to distinguish between them since it would be bad if a person requests the dialog agent to start the music and they did the opposite. However, antonyms can actually be confused with each other in embedding models like Word2Vec since they are typically one of the words in the embedding space that is most similar to a term.

4.2.2 WordNet: A Database of Lexical Relations

The WordNet is a lexical database with a lot of different words and sentences that are used in the Natural Language Processing. The lexical relationship is framing a proper sentence with the help of the different combination of words. For example, "I ride a Bicycle." Is different from "I was riding a Bicycle." The lexical relationship in the example is bicycle. Bicycles are entities. I is an actor. All the information is correctly mapped inside the WordNet. The WordNet also contains information like noun, verb, adjective, and adverb forms. The WordNet stores sense relations. The sense relations are:

1. Hypernym: It is a relationship between concept and its superordinate
   Example: Food is a hypernym of cake.
2. Hyponym: It is a relationship between concept and its subordinate.
   Example: Cargo is a hyponym of dog.
3. Holonym: It is a relationship between a whole and its parts.
   Example: Lorry is a holonym of a wheel.
4. Meronym: It is a relationship between part and its whole.
   Example: Wheel is a meronym of lorry.

WSD is used to identify the meaning of a word which can be used in a different scenario. The first problem which the natural language processing faces is lexical ambiguity, syntactic or semantic. The part of speech tagging with maximum accuracy will provide solution to the word's syntactic disambiguity. In order to solve the semantic ambiguity, the word sense disambiguation is used. The dictionary and the test corpus are the two inputs used to evaluate the WSD. In Figure 4.1, the WSD is derived from non-predefined senses and predefined senses. The predefined senses are acquired from knowledge and corpus.

4.2.3 Approaches to Word Sense Disambiguation

Different types of WSD are explained here:

- Dictionary based

The dictionary base is the primary source for WSD. The corpus is not used to solve the disambiguation. The Lesk algorithm was developed by Michael Lesk in 1986. It is used to remove the word ambiguity.

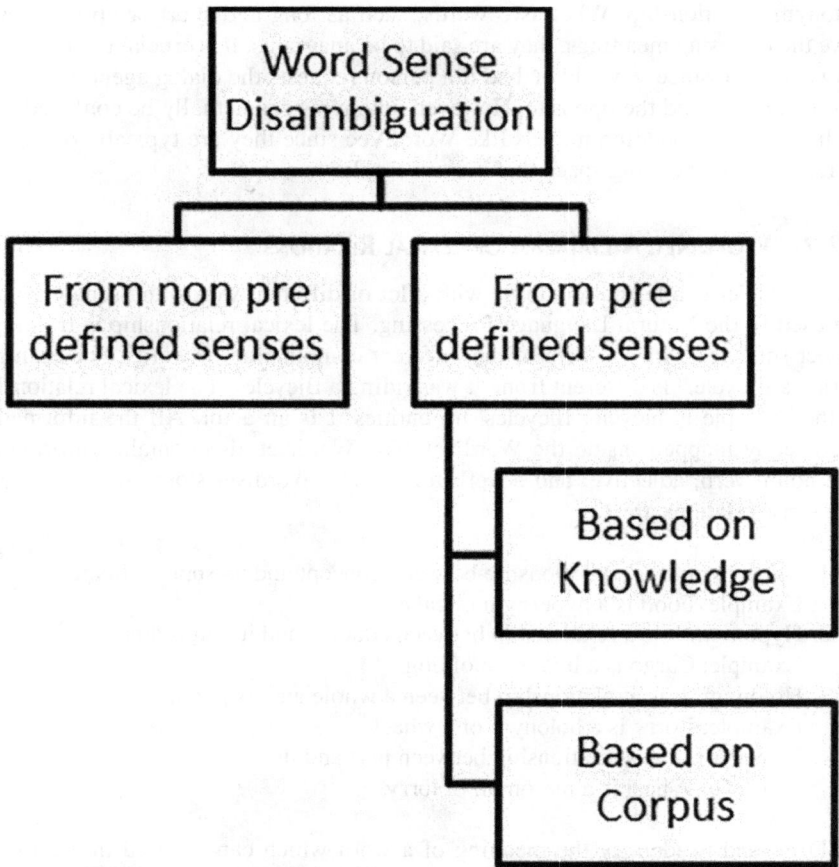

FIGURE 4.1   Word Sense Disambiguation.

- Supervised methods

Machine learning approaches use sense-annotated corpora to train for disambiguation. The terms "knowledge" and "reasoning" aren't used in these procedures. The context is represented by a set of "word characteristics." It also contains information about the words that surround it. The most prominent supervised learning approaches to WSD are support vector machine and memory-based learning. These approaches rely on a large number of manually sense-tagged corpora, which are time-consuming and costly to produce.

- Semisupervised methods

The word disambiguation algorithms are mostly semi-supervised. Since they are semi-supervised, they use both labeled and unlabeled data. This particular method needs small amount of annotated text and large amount of plain unannotated text.

• Unsupervised methods

In unsupervised methods, similar senses occur in similar context. The senses can be inferred by word grouping the number of times the word occurs, and it is also based on the same measure of context similarity. Due to their lack of reliance on manual efforts, unsupervised approaches have a lot of potential for overcoming the knowledge acquisition barrier.

4.2.4 APPLICATIONS OF WORD SENSE DISAMBIGUATION

Some of the applications of WSD are machine translation, information retrieval, and text mining as represented in the Figure 4.2.

• Machine translation (MT)

Machine translation is basically converting one human language to another human language without any human involvement. One of the popular machine translation tools is Google Translator. The Google Translator is widely used software by Google. Google Translator is very useful while travelling. The machine translation is widely used in WSD. WSD makes the lexical choices in Machine Translation for terms that have different translations for related meanings. The target language uses terms to correspond to the senses in Machine Translation. The vast majority of machine translation systems don't employ an explicit WSD module.

• Information retrieval (IR)

Information retrieval is the practice of organizing, storing, retrieving, and evaluating information from database, which consists of documents, which are mostly text-based documents. This technology is used to help the users in order to locate data, but it will not return the exact answers directly. The ambiguities in the queries submitted to the IR system are resolved using WSD. Similar to MT, contemporary IR systems rely on the user providing sufficient context in the query to only return pertinent documents rather than expressly using WSD module.

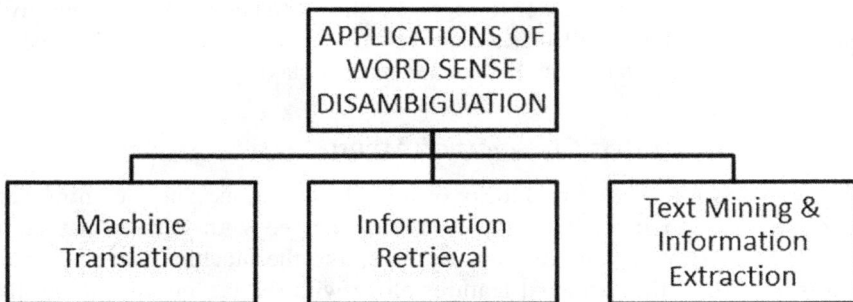

FIGURE 4.2 Applications of Word Sense Disambiguation.

- Text mining

Text mining is the concept of analyzing large amount of unstructured text data and involves using the software to identify the insights within the text, patterns, keywords, and various properties in the data. The alternate name for text mining is text analytics. Text mining is frequently used by data scientists and others for developing the big data and deep learning algorithms. The chatbots and virtual agents can be seen nowadays in most of the webpages. These applications are used to acquire texts. The word sense disambiguation is widely used to do relevant analysis of text. The WSD is used to find the correct words.

4.3 TEXT CLASSIFICATION

Text classification is the process of selecting one or more types of text from a wider pool of potential categories to place a given text. For example, the majority of email service providers have the beneficial function of automatically separating spam emails from other emails. This is an example of how text classification, a common NLP problem, is used. Each incoming email is classified into one of two categories in the email spam-identifier example: spam or non-spam. There are many different applications for this process of classifying texts on the basis of certain features in different fields, including social media, e-commerce, healthcare, legal, and marketing. Although the goal and use of text classification may differ from one domain to another, the underlying problem remains the same.

A specific case of the classification problem is text classification, where the objective is to classify the input text into one or more buckets (referred to as classes) from a collection of predefined buckets (classes). Any length of "text" may be used, whether a single character, a word, a sentence, a paragraph, or an entire document. Imagine that the user wishes to divide all customer reviews for a product into three groups: favorable, negative, and neutral. The difficulty of text categorization is to "learn" this categorization from a set of samples for each of these categories and forecast the categories for fresh, undiscovered goods and fresh customer reviews.

The text classification is divided into three types, which are binary, multiclass, and multi-label. In order to check the email is spam or not spam, it is termed as binary classification; and in order to segregate the customer feedback as positive, negative, and neutral, it is called as a multiclass classification. In multilabel classification, the document can have more than one label or class.

4.3.1 BUILDING THE TEXT CLASSIFICATION MODEL

Figure 4.3 shows the building blocks of text classification. The incoming text is considered as an input, the input text is processed with feature extraction method, after extracting the relevant features, and the machine learning algorithm is applied to the extracted features along with the tag in order to get the final model.

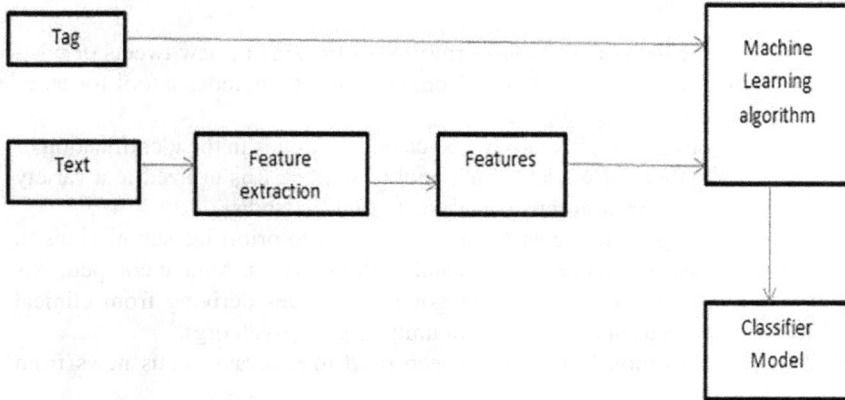

FIGURE 4.3 Building Blocks of Text Classification.

The steps involved are:

1. Collect the dataset which has corresponding labels.
2. The dataset is divided into training set and testing set.
3. Convert the raw data into feature vectors.
4. Employ the feature vectors from the training set along with the relevant labels to train a classifier.
5. Evaluate the performance metrics.
6. Deploy the model.

4.3.2 Applications of Text Classification

* Customer support

Social media is a popular platform for consumers to share their thoughts and experiences with goods and services. Text classification is frequently used to distinguish between tweets that require a response from brands (those that are actionable) and those that do not require any response.

* E-commerce

On e-commerce sites like Amazon and eBay, customers post reviews for a variety of products. To comprehend and analyze customers' perceptions of a product or service on the basis of their remarks is an example of how text categorization is used in this type of scenario. This practice is referred to as "sentiment analysis." It is widely used by brands all over the world to determine if they are drawing nearer to or further away from their consumers. Over time, sentiment analysis has developed into a more complex paradigm known as "aspect"-based sentiment analysis, which classifies consumer input as "aspects" rather than merely positive, negative, or neutral.

4.3.3 OTHER APPLICATIONS

- Text classification is used to determine the language of new tweets or postings, for example. For instance, Google Translate includes a tool for automatically identifying languages.
- Another common application of text categorization is in the identification of unknown authors of works from a pool of writers. It is utilized in a variety of disciplines—from forensic analysis to literary studies.
- In the recent past, text classification was used to prioritize submissions in an online discussion board for mental health services. Annual competitions for the solution of such text categorization issues deriving from clinical research are held in the NLP community (e.g., clpsych.org).
- Text categorization has recently been used to separate bogus news from actual news.

4.4 SENTIMENT ANALYSIS

Sentiment analysis is otherwise called opinion mining in Natural Language Processing. It is a concept of Natural Language Processing used to recognize emotion. Sentiment Analysis is one of the techniques used by different organizations to identify and organize thoughts about the products, services, and customer feedbacks. Sentiment Analysis uses the data mining concept, machine learning, and AI in order to mine the text for sentiment and subjective information.

Sentiment Analysis acts as a tool in businesses for extracting information from unstructured text taken from blogs, reviews, chats, social media channels, and advertisements. Apart from extracting the sentiment, opinion mining will identify whether the given text is positive, negative, and neutral. The third-party vendors available in the market are Netbase, Lexalytics, and Zoho provided by Software as a Service (SAAS) model.

Sentiment analysis is otherwise a contextual text mining that recognizes and takes the subjective information from the source content. It is very useful in online discussions, especially in business sector by understanding the brands, products, and services. However, simple sentiment analysis and count-based metrics are used in social media stream analysis. For example, recent developments in deep learning have greatly increased algorithms' capacity for text analysis. The inventive application of cutting-edge AI methods can be a useful instrument for doing a deep study.

Figure 4.4 shows the generic framework of Sentiment Analysis system. From the corpus, the documents are processed, after which, the analysis is made. Once the documents are analyzed, the sentiment scores for entities and aspects are performed. The document analysis is also referred to lexicons and linguistic resources.

4.4.1 TYPES OF SENTIMENT ANALYSIS

The variants of sentiment analysis are described in detail are as follows:

- Graded sentiment analysis

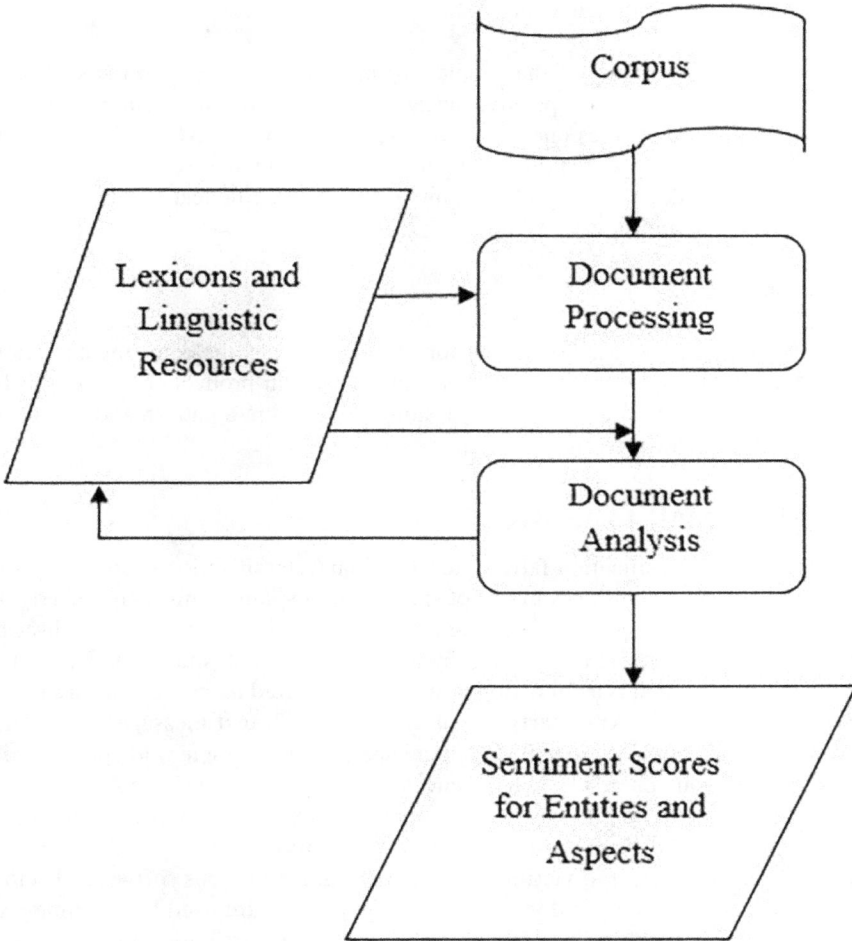

FIGURE 4.4 Architecture of Generic Sentiment Analysis System.

The polarity plays an important aspect in Graded Sentiment Analysis. The polarity can be expanded into different levels of positive and negative. The ratings can be given as 5 (for very positive) and 1 (for very negative). This is based on the customer feedback.

- Emotion detection

This approach of reading a text's emotions is trickier. Machine learning and lexicons are used to determine the sentiment. Lexicons are collections of positive or negative word lists. This makes it easier to classify the terms according to their usage. The advantage of doing this is that a company can understand why a consumer feels the way they do. This is more algorithm-based and may be challenging to comprehend at first.

- Aspect-based sentiment analysis

The specific characteristics of the people will be discovered through the sentiment analysis and are referred to as positive, negative, and neutral. This form of sentiment analysis often just examines one aspect of a service or product. To understand how customers feel about specific product attributes, a company that sells televisions, for example, may use this type of sentiment analysis for a specific feature of televisions, such as brightness and sound.

- Intent analysis

This analysis is purely based on the customer. For example, the company can easily predict whether the customer will purchase this particular product or not. This is by tracking the intention of the particular customer Producing a pattern and then used for marketing and advertising.

4.5   SPAM EMAIL CLASSIFICATION

The spam email classification falls under the binary classification problem. It is to check whether the email is spam or not spam. Email spam, commonly referred to as junk email, is the practice of sending unwanted email to a large list of subscribers in mass. Spam can be transmitted by actual people, but, more frequently it is sent by a botnet, which is a collection of malware-infected machines that are under the control of one attacking party. In addition to email, text messages and social media platforms can also be used to spread spam. Many people find spam emails annoying. The spam emails need to be checked regularly, and you should delete the spam emails very frequently, as if not properly checked, they will clog the inbox mails. Spammers that send spams via email frequently change their strategies and content to deceive potential victims into downloading malicious software, disclosing personal information, or donating money. Spambots are used by spammers to search the Internet for email addresses to add to email distribution lists. The lists are used to simultaneously distribute spam to thousands or even millions of email addresses.

Spammers can simply send their dodgy message to a large number of email addresses in one go, even though they only want a tiny percentage of recipients to respond to or interact with their message. Because of this, spam is still a significant issue in the contemporary digital economy.

The spam subjects mostly are of the form:

- Medical field
- Financial services
- Online courses
- Work from home jobs
- Online games
- Online gambling
- Cryptocurrencies

4.5.1 HISTORY OF SPAM

Although spam may be a contemporary issue, it has a lengthy history. Gary Thuerk, a worker at the now-defunct Digital Equipment Corp. (DEC), wrote the first spam email in 1978 to advertise a fresh item. There are 2,600 users who had email accounts on the Advanced Research Projects Agency Network, and the unwanted email was sent to around 400 of them. According to some accounts, DEC's new sales increased by around $12 million as a result.

However, the term "spam" didn't come into use until 1993. It was used with Usenet, a newsgroup that combines elements of both email and a website forum. It posted more than 200 messages to a discussion group automatically due to a bug in its new moderation software. The event was mockingly referred to as spam.

In 1994, Usenet was also a target of the first significant spam attack. Spam accounted for 80–85% of all emails sent globally in 2003. The United States Passed the Controlling the Assault of Non-Solicited Pornography and Marketing (CAN-SPAM) Act of 2003 as a result of the problem becoming so pervasive. The most crucial law that legal email marketers must abide by to avoid being branded as spammers is still CAN-SPAM.

The volume of spam sent on a daily average decreased from 316.39 billion to roughly 122 billion between mid-2020 and early 2021. However, spam still makes up for 85% of all emails, costing reputable companies billions of dollars annually.

4.5.2 SPAMMING TECHNIQUES

- Botnets: Using command-and-control servers to gather email addresses and disseminate spam is made possible by botnets.
- Blank email spam: This method entails sending emails with a blank subject line and message body. By finding incorrect bounced addresses, it might be used in a directory harvest attack to validate email addresses. In certain cases, emails that appear to be empty may really include viruses and worms that can propagate via embedded HTML code.
- Image spam: The email body contains a JPEG or GIF file that contains the message content, which is computer-generated and unreadable by human users. This approach makes an effort to evade detection by text-based spam filters.

4.5.3 TYPES OF SPAMS

- Malware messages: Few spam emails contain malware, which can fool users by disclosing the personal information, making payments, or taking some action.
- Frauds and scams: Users get emails with offers that promise rewards in exchange for a small deposit or advance charge. Once they have paid, the fraudsters will either create new charges or cease communication.
- Antivirus warnings: These notifications "warn" a user of a virus infection and provide a "fix" for it. The hacker can access the user's system if they fall

for the trick and click on a link in the email. A malicious file could also be downloaded to the device via the email.

- Sweepstakes' winners: Spammers send emails with the false claim that the receiver has won a contest or prize. The email's link must be clicked by the receiver in order to claim the prize. The malicious link usually seeks to steal the user's personal data.

4.6 QUESTION ANSWERING

In this digital era, the volume of data is rapidly growing on a day-by-day basis. The expected data on 2025 would be 181 zettabytes. The data would be either structured or unstructured. The data may be in the form of blogs, documents, messages, voice messages, or videos. Finding the answer for all the questions is neither easy nor tough. Thus, a new approach has been introduced called Question Answering.

Question answering is one of the main applications of Natural Language processing. It is built in such a way that it automatically answers queries posted by humans. Question Answering has got a rich history, and it is now very popular in Computer Science and Natural Language Processing. Question answering is a significant NLP problem as well as a long-standing AI milestone. A user can ask a question in natural language and receive a rapid and concise response using QA systems. Search engines and phone conversational interfaces now have quality assurance processes, and they're really good at answering simple questions. On more difficult inquiries, however, these typically merely offer a list of snippets for us, the users, to browse through in order to locate the answer to our inquiry.

4.6.1 COMPONENTS OF QUESTION ANSWERING SYSTEM

Figure 4.5 shows the question answering system. The retriever and reader are the basic components of most modern extractive natural language question answering systems. The system tries to discover and extract a suitable answer to a user's question from a multitude of texts rather than keeping the answers to the questions in massive databases. First, documents from a document store that are pertinent to the user's query are loaded. The reader then tries to deduce the response to the user's query.

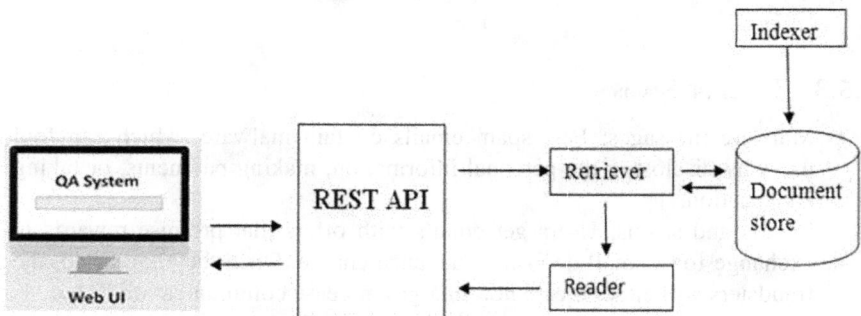

FIGURE 4.5 Question Answering System.

4.6.1.1   Document Store

The document store is used to store the relevant documents. There are a number of techniques to retrieve the documents from the store. The most common technique used is simple or reverse index. The Apache Lucene is a reverse index used as an index in Elastic Search. It is one of the fastest and efficient ways for retrieving documents from the document store.

4.6.1.2   Retriever

The task of a retriever is to locate pertinent documents for the user's inquiry. It tries to extract the pertinent terms from the question first. Then it uses these to find pertinent materials. Several Natural Language Processing (NLP) approaches are utilized to transform a user's inquiry into a form that a retriever can understand. These consist of:

- Removing punctuations

When finding pertinent materials, full stops, commas, and other punctuations are redundant. As a result, they are eliminated from the user's query.

- Removing stop words

Stop words are often used words that don't significantly change the meaning of the text. Examples are articles such as the, a, and an. As a result, these words are eliminated.

- Tagging entities

Entities that are directly related to the query are typically things like products or names. As a result, they are included in the query.

- Stemming

Words can take on several guises or conjugations (walk, walked, walking, etc.). Such terms are stripped down to their most basic form before being put into the query because they may well appear in many forms within a document.

4.6.1.3   Reader

The responsibility of the reader is to extract an answer from the documents they receive. They make an effort to comprehend the question and the documents by using a suitable language model, and then they mine the texts for the best possible response.

4.6.1.4   Language Model

Language models come in numerous varieties. Depending on the fields in which they are used, their results will vary in quality. These models can understand and process language since many of them use a transformer or attention-based approach. The BERT is a language model mostly used in the question answering system, which

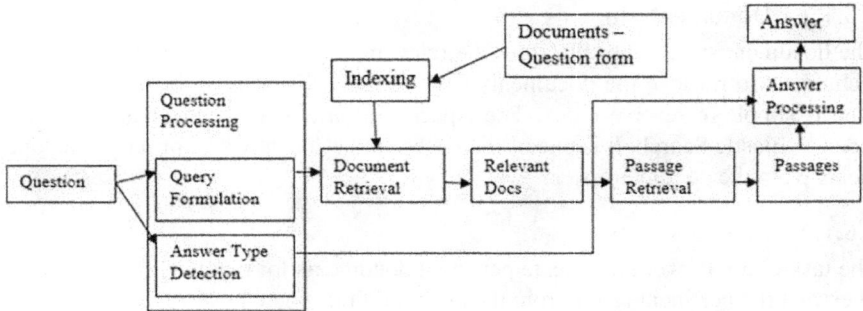

FIGURE 4.6   Information Retrieval-based Factoid Question and Answering System.

uses transformers or attention-based models that are used to develop the question answering system.

4.6.2 Information Retrieval-based Factoid Question and Answering

In Figure 4.6, the process flow of the information retrieval system is depicted. The IR-based factoid question and answering has three parts, and they are question processing, passage retrieval, and answer processing.

- Question processing

In IR-based factoid question answering, the first step is to retrieve the query. Initially, the keywords are extracted; next, the entity type or answer type is gathered. The focus (word replaced by the answer) question types and relations are extracted.

- Answer type detection

Answer type detection is a classification task based on supervised learning. The answer types can be constructed automatically from the sources like WordNet.

- Query formulation

The query formulation is the process of passing a question to the web search engine The question answering from the smaller sets like corporate information pages and more processing such as query expansion and query reformulation are performed.

- Passage retrieval

One of the basic methods of passage retrieval is to send every paragraph to the answer extraction stage. One of the more advanced techniques is that to filter the paragraph after executing the named entity or answer type classifications on the passages.

The ranks can be completed with the help of the supervised learning using features such as:

1. The named entities in the passage
2. The total number of question-related terms in the passage
3. The rank of the document

- Answer processing

The answer processing is the final step in the question answering section. The final step is to extract the relevant answer from the passage.

4.6.3 ENTITY LINKING

The entity linking is the process of providing the distinct identification to entities like location. Entity linking can initiate question answering and combining the information. Steps involved in entity linking are as follows:

- Information extraction: Retrieve information from the unstructured data.
- Named Entity Recognition (NER): Organizations and places are some of the examples of the Named Entity Recognition. In text, NER recognizes and categorizes the occurrences of named entities into predefined categories. NER is the role of assigning a tag to each word in a phrase.
- Named Entity Linking (NEL): NEL will assign a unique identity to each entity recognized by NER. NEL then tries to link each entity to its knowledge base description. The knowledge base to be used depends on the software, however, for open-domain literature; Wikipedia-derived knowledge base like Wikidata, Dbpedia, or YAGO could be used.
- Entity linking can be done in one of two ways: end-to-end or end-to-end with recognition and disambiguation. Entity linking only provides disambiguation if gold-standard-named entities are accessible at the input.

Steps involved in implementing an entity linking system are as follows:

- Recognize

Recognize items referenced in the context of the text. The entity linking mechanism in this module seeks to remove unwanted entities in the knowledge base. For every entity mention m and return, a candidate entity set Em consists of relevant entities.

- Rank

Assign a score to each candidate. The size of the candidate entity set Em is usually greater than 1. To rank the potential entities in Em, researchers use many types of evidence. They try to locate the entity Em, which is the most plausible link for the reference m.

- Link

In the knowledge graph, connect the recognized entities to the categorized entities.

4.6.4 Knowledge-based Question Answering

The goal of knowledge-based question answering (KBQA) is to respond to a natural language question using a knowledge base (KB) as the source of information. The knowledge base is a structured database. It contains the collection of facts in the form (subject, relation, and object) where each fact will have properties attached to it as qualifiers.

Example: "Rajiv Gandhi got married to Sonia Gandhi on 25 February 1968 at wedding hall." It can be represented in the tuple as <Rajiv Gandhi, Spouse, Sonia Gandhi') with the qualifiers start time = 25 February 1968 and place of marriage= wedding hall.

Types of knowledge-based question answering:

- Simple question answering involves single fact.
- Complex question answering involves multi-hop reasoning, numerical operations, constrained relations, or combinations of the same.

4.6.4.1 Approaches of Knowledge-based Question Answering

Answering the simple question is easy and straightforward. Whereas complex questions are typically difficult to answer because complex question answering involves multi-hop reasoning over the knowledge base, rules, and operations. There are two approaches available for the complex question answering, they are:

- Semantic parsing-based method (SP-based method)

The semantic parsing-based approach works on the principle of parse-then-execute principle. The goal of this technique category is to parse a natural language utterance into logic forms. In order to predict answers, the semantic parsing-based approach follows the steps given here:

- Parse the natural language question into an uninstantiated logic form (e.g., a SPARQL query template), which is a syntactic representation of the question free of entities and relations.
- The logic form is subsequently instantiated and validated by using KB grounding to perform various semantic alignments to structured KBs (obtaining, e.g., an executable SPARQL query).
- To create expected replies, the parsed logic form is run against KBs.
- Information retrieval-based approaches.

The information retrieval-based approach works under the principle of retrieval and rank. It follows the steps given here:

- The system first extracts a question-specific graph from KBs, ideally comprising all question-related entities and relations as nodes and edges, starting with the topic entity.
- The system then converts the input questions into vectors that convey reasoning instructions.

- A graph-based reasoning module performs semantic matching using vector-based computation to propagate and aggregate data across the graph's nearby entities.
- The entities in the graph are ranked according to their reasoning state at the end of the reasoning phase using an answer-ranking module. The answers to the question are projected to be the top-ranked entities.

Advantages and disadvantages of semantic parsing-based approach and IR-based approach are as follows:

Semantic parsing-based methods offer a more interpretable reasoning process. Their strong reliance on the design of logic forms and parser modules is used primarily to improve the performance.

IR-based solutions are easy to train because they readily fit into popular end-to-end training.

4.7   CHATBOTS AND DIALOG SYSTEMS

The conversational agents and dialog systems will interact with users in natural language either in text, speech, or both. The task-oriented dialog agents will communicate with users in order to complete the task. Dialog agents such as Siri and Google Talk provide assistance to control applications or to make calls. For example, the conversational agents will identify that the user is parking in the correct parking area, else the parking fine will be intimated to the user. The chatbots are designed in such a way so that they can aid in extended conversations, setting up an unstructured conversation of human–human interaction.

4.7.1   PROPERTIES OF HUMAN CONVERSATION

Human conversation is a nuanced and complex shared endeavor. It is imperative to comprehend some aspects of human communication before attempting to develop a conversational bot that can interact with humans. The properties of human conversation are:

- Turns: Turns are separate contributions to the dialog. Turns will be in the form of a single word (i.e., in a shorter sentence) or with multiple words (in a longer sentence). It is very important to understand the structure of the turn for a spoken dialog system. The system has to decide when to start talking and stop talking. The system needs to detect as soon as the user finishes the conversation called endpoint detection.
- Speech acts: The speech acts referred to the actions are performed by the speaker. The speech acts are otherwise called dialog acts.
- Constatives: Constatives are creating a statement—for example, answering the question. Some of the constatives are answering, claiming, and confirming.
- Directives: A directive is an act in which you try to get your conversational partner to do something. The directives may be advising someone, asking them to do something, provide some kind of information, forbidding,

inviting them, ordering them to do some work, and politely requesting them to do some work.

- Commissives: Commissives are where commitments are made—a kind of a future action. Commissives can be making plans, explicit promising, vowing to do something, betting or showing explicit opposition.
- Acknowledgments: Acknowledgments provide a useful function by expressing a speaker's attitude toward some sort of social action. Examples of acknowledgments include apologizing, greeting, expression of gratitude, and acceptance.
- Grounding: Grounding is establishing the common ground between two parties in the conversation. It is to acknowledge that the speaker has been heard or understood. This is usually by saying okay at the beginning of the turn, repeating the parts of what the other speaker said, and using other implicit signals
- Initiative: The initiatives are conversational controls. The speaker asking questions has a conversational initiative. In every dialog, most interactions are mixed initiative. Initiative is a sense of control in the conversation. Even though most human–human conversations are mixed initiative, it is very difficult for the dialog systems to achieve mixed initiative conversations.
- Structure: Conversations have structure. Questions set up an expectation for an answer, and proposals set up an expectation for an acceptance or rejection.
- Adjacency pairs: The dialog system's act pairs are adjacent pairs.
- Inference: Providing conclusions based on more information than is present in the uttered word.
- Implicature: The act of implying meaning beyond what is directly communicated.

4.7.2 Chatbots

NLP enables your chatbot to evaluate and generate text from human language. NLP (natural language processing) is an AI tool that aids your chatbot in analyzing and comprehending natural human language exchanged with your clients. Chatbots may grasp the conversation's intent rather than just using the data to communicate and reply to questions. In the area of automation and AI, there are various acronyms that are important to know in order to understand how your virtual agent or NLP chatbot operates. They are NLU, natural language generation (NLG), and natural language interaction (NLI).

4.7.2.1 Natural Language Understanding (NLU)

Natural language understanding (NLU) is a type of an AI. With the help of AI, it uses computer software to comprehend input in the form of spoken or in written format. Human–computer interaction is made possible via NLU. Computers can comprehend commands without the codified syntax of computer languages since they can understand human language, such as English, Spanish, and French. NLU's major goal is to develop chat- and voice-capable bots that can communicate with the public without any help. NLU projects are now being worked on by many well-known IT firms, including startups and global leaders like Amazon, Apple, Google, and Microsoft.

4.7.2.2 Natural Language Generation (NLG)

In order to create meaningful sentences in natural language, natural language generation is used. It induces writing in computers. Based on the input data, NLG is utilized to generate a text answer in human language. Text-to-speech services can also turn the provided text into speech. NLG may summarize texts as well.

4.7.2.3 Natural Language Interaction (NLI)

Using NLI, the machine will be able to communicate with people using their own language when the data is entered in their own language and the machine will respond in a comprehensible manner.

4.7.2.4 Types of Chatbots

The various chatbots available are discussed in this section

• Menu-button-based chatbots

The most fundamental form of chatbots now used on the market is menu/button-based ones. These chatbots are typically based on the decision tree hierarchies that appear to the user as buttons. These chatbots demand the user to make a number of decisions in order to develop deeper and get at the ultimate solution, much like the automated phone menus.

• Linguistic-based chatbots

These chatbots are used to predict the type of questions the customer may ask. It creates the conversational automation logic. As the first step, language conditions need to be defined clearly. The conditions will be of the form to assess the word, order of the word, and context of the word.

• Keyword-recognition-based chatbots

This chatbot listens to the customer input, and the input will be typed by the customer. It recognizes the customer input and it will reply accordingly. This chatbot is developed using the artificial intelligence concept.

4.7.2.5 Working of Chatbot

There are five major steps involved in the working of the chatbot, they are tokenizing, normalizing, recognizing entities, dependency parsing, and generation.

• Tokenizing: The chatbot begins by breaking up text into small chunks (also known as "tokens") and deleting punctuation marks.
• Normalizing: The bot then removes irrelevant information and changes words to their "regular" form, such as by making everything lowercase.
• Recognizing entities: Now that all of the words have been normalized, the chatbot tries to figure out what kind of thing is being discussed.

- Dependency: The bot then determines the function of each word in the sentence, such as noun, verb, adjective, or object.
- Generation: Finally, the chatbot develops a number of responses on the basis of the data gathered in the previous steps and chooses the most appropriate one to send to the user.

4.7.3 THE DIALOG-STATE ARCHITECTURE

A more advanced variation of the frame-based architecture is known as the dialog-state system, which serves as the foundation for contemporary research systems for task-based discussion. In Figure 4.7, there are six components of a typical dialog-state system.

- Automatic speech recognition

An audio input can come from a phone or any other device, and the output of the ASR will be a continuous string of words. Similar to a discussion state, the ASR component is likewise. For instance, what if the computer asked the user which state they were leaving? The ASR model will respond with state names with a high probability under certain situations. The language model is trained to accomplish the same.

- Spoken language understanding

The speech synthesis component is the key component for the spoken language processing. In the dialog state architecture, has component for retrieving the slot fillers from the users input, with the help of machine learning rules.

- Dialog state tracking

The conversation state tracker used in the architecture is used to retrieve the users' present state of the frame and users' most recent conversation. The dialog state

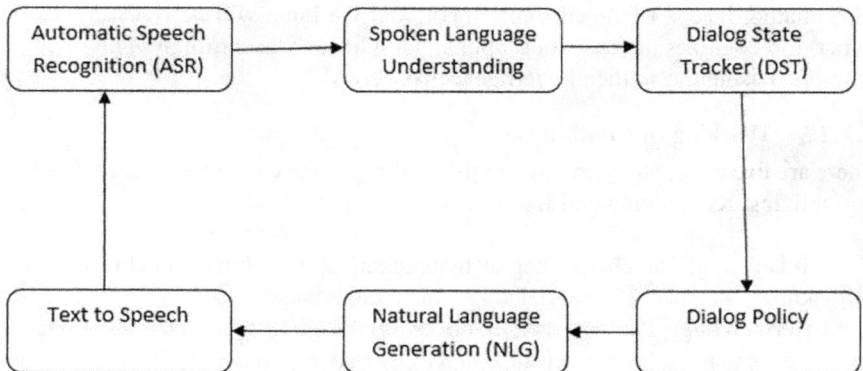

FIGURE 4.7   Architecture of the Dialog-state System.

encompasses current sentence's slot-fillers and also complete state of the frame. It means that it collects all the users' limitations.

- The dialog policy

The dialog policy's purpose is to determine the generation of action and dialog act. In more technical terms, it needed to be forecasted which action A_i will be done at turn i in the conversation based on the overall dialog state. The state refers to the complete series of dialog acts between the user and the system. It is referred for User (U) and for system (A).

$$A_i = \underset{A_i \varepsilon A}{\arg\max} \, P\left(A_i \# A_1 U_1 \ldots \ldots A_{i-1}, U_{i-1}\right) \qquad (1)$$

- Natural language generation (NLG)

As soon as the dialog act has been decided, the response to the user query needs to be generated in the form of text. In the information state architecture, the NLG is divided into two types: they are content planning (what to say) and sentence realization (how to say). It is assumed that the content planning is done by the dialog policy.

- Text to speech

After the Natural Language Generation step, the text is converted into speech.

4.8   SUMMARY

The Natural Language Processing has been emerging steadily in the recent years. It is used in many organizations and industrial applications. This chapter gives the idea about the recent and traditional applications of Natural Language Processing. It explains most popular applications like WSD, which is mainly used in text mining and the tasks involving the extracting the information. Word Sense Induction is used in Web search clustering. Text Classification is used in social media, customer experience, marketing, and in many more fields. Sentiment analysis is mainly used in customer service management and for analyzing the customer feedback. Question Answering system is used in IBM Watson. Spam email classification is primarily used to classify spam or not spam. Information retrieval is used in digital libraries' blog search, chatbots and dialog system, and properties of human conversation. They are the current research trends which researchers can explore using the state-of-art-learning methods, providing good venues for research and problem statement for research in NLP.

BIBLIOGRAPHY

Christopher D. Manning and Hinrich Schütze, *Foundations of Statistical Natural Language Processing*, Cambridge: The MIT Press, 2018.

Daniel Jurafsky and James H Martin, *Speech and Language Processing: An introduction to Natural Language Processing, Computational Linguistics and Speech Recognition.* Hoboken, NJ: Prentice Hall, 2014.

Ela Kumar, *Natural Language Processing.* New Delhi: IK International Pvt Ltd, 2011.

James Allen, *Natural Language Understanding.* Benjamin: Cummings Publishing Company, 2003.

Li Deng and Yang Liu, *Deep Learning in Natural Language Processing.* Berlin: Springer, 2018.

Madeleine Bates and Ralph M. Weischedel. *Challenges in Natural Language Processing.* Cambridge: Cambridge University Press, 2006.

Nitin Indurkhya and Fred J. Damerau, *Handbook of Natural Language Processing Machine Learning & Pattern Recognition Series.* London: Chapman & Hall/CRC, Taylor and Francis Group, 2010.

Steven Bird, Ewan Klein, and Edward Loper, *Natural Language Processing with Python—Analyzing Text with the Natural Language Toolkit.* London: O'Reilly, 2012.

Tanveer Siddiqui, *U.S. Tiwary, Natural Language Processing and Information Retrieval.* Oxford: Oxford University Press, 2008.

Yoav Goldberg, *Neural Network Methods for Natural Language Processing.* London: Synthesis Lectures on Human Language Technologies, 2017.

5 Fundamentals of Speech Recognition

LEARNING OUTCOMES

After reading this chapter, you will be able to:

- Understand the fundamental theory of speech recognition.
- Know the basics of speech and its characteristics.
- Identify the different models and problems/challenges when designing ASR.

5.1 INTRODUCTION

In today's computer and mobile era, speech has become a more important channel for human–machine connection. Human–computer interaction has been evolving since the dawn of computer engineering. In today's world, it is not uncommon for this engagement to take place through speech. Several software packages now incorporate cutting-edge speech technology to perform a variety of tasks. A detailed study of human speech perception is required for these systems to be of practical utility, i.e., to perform in a human-like manner. A compact and meaningful representation of speech input, which eliminates the influence of inconsequential components like as background noise, is also a key factor in improving the system's performance. Speech input has recently started to change the way people interact with one another. This method of communication is very useful in several applications like assistive technology for disabled people, autonomous vehicle for getting the navigation, and multimedia search. ASR technique aims in making the computer understand human speech and respond. Given the speech signal, ASR technology derives the transcribed utterances. The fundamental difference between speech recognition performed by people and computers using automatic voice recognition is given here:

- As humans, we are able to differentiate between the speech variations and inconsistencies, as well as the ability to differentiate between /a/ and /e/ sounds made by various speakers and uttered in various situations and languages, etc. But computers need to be taught and made to learn which the invariant features are e.g.: /a/ or a /e/ sound.
- The second problem is that speech is a continuous signal, meaning that humans don't speak in single words but rather continuously. Actually that is what we call the effect of co-articulation which means that one sound is linked to the other sound, and the movements of the articulatory organs span over from one sound to the next and to the following one.

DOI: 10.1201/9781003348689-5

It is a challenge for a computer to understand continuous speech.

The history of ASR starts over last 100 years. The recent advancements in spectral analysis, cestrum, dynamic time wrap, HMM, language model, and deep learning models attained good performance nearer to human recognition. It is observed that a committed improvement is achieved over the past years. The major milestones in ASR are:

1922: Radio-Rex speech recognition system.
1939: Voder and vocoder.
1952: Isolated one-digit recognizer from Bell Labs.
1957: Olson and Belar of RCA laboratory developed ten syllables of a solo voice.
1980: Worlds of Wonders Julie Doll.
1990s: Dragon Dcatate.
2000s: Voice search is a service provided by Google.
2010s: Virtual assistants include Siri, Cortana, Alexa, and Google Assistant.

5.2   STRUCTURE OF SPEECH

Speech is produced by the change in air pressure that produces sound wave that our ears process with the support of our brain. Speech is a complex phenomenon, and it is represented as a result of continuous articulator movements. Speech gets originated in the vocal tract, and various articulator movements of tongue, teeth, and the waves are modulated by lips. Speech is produced when air waves exit through the mouth and nose. Human speech varies in a range of 85 Hz to 8 kHz, and the human hearing range is 2 Hz to 20 kHz. Words make up speech, and each word is made up of phones. Speech is a dynamic process with no clearly defined elements since there are no limits to the number of units or words that can be used.

Figure 5.1 represents the audio stream of continuous speech where the stable states mix with dynamically changed state. The acoustic feature of a waveform correlates to a phone, which is determined by a variety of characteristics such as the phone environment, speaker, and speech style. Two consecutive phones are called diphone. When phones are considered in context, they are often called as triphone or quinphone. Phones build subword units called syllables. When we speak spontaneously, phone numbers vary frequently, but the syllables remain the same. There are no limits to the number of units or words that can be used. Words are vital in voice recognition system because they limit the number of phone combinations that can be used. There must be 40 to the power of seven words if there are 40 phones, and the average word has seven phones. Even persons with a large vocabulary rarely use more than 20,000 words in their daily lives, making recognition much easier. Words and other non-linguistic noises such as (breathe, um, uh, and cough) are referred to as "fillers" in utterances.

5.3   BASIC AUDIO FEATURES

Before using feature extraction algorithms, the audio signals are separated into frames. Initially, all the values corresponding to features are obtained. The long-term features of an audio signal are then determined using a texture window functions. In the next sections, some of the most commonly utilized audio features are discussed.

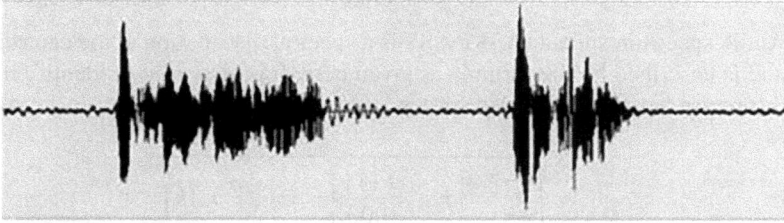

FIGURE 5.1 Audio Stream of Continuous Speech.

5.3.1 PITCH

Vocal cord vibrates only when a sound is produced, which in turn generates glottal pulse. Pitch is the glottal pulse's fundamental frequency. It identifies a specific tone and distinguishes between different sounds. Time frequency domain can be used to analyze pitch. Zero crossing method can be used to determine pitch.

5.3.2 TIMBRAL FEATURES

The speech signals have same pitch and loudness differentiated by the timbral features. They represent the sound quality. The harmonic component of an audio determines the timber. The vibrato and tremolo present in speech also determine timbre. The dynamic characteristics of speech are vibrato which increases the richness of speech.

5.3.3 RHYTHMIC FEATURES

The rhythmic elements of an audio signal, which are of two types, rhythmical structure and bit strength, determine the regularity of the signal. Histogram is used to represent the bit strength.

5.3.4 MPEG-7 FEATURES

The Moving Pictures Expert Group (MPEG) has established an international standard for the classification of audio/speech signals. It defines the following audio feature standards.

- Audio spectrum centroid (ASC): Audio Spectrum Centroid (ASC) is standardization in in MPEG-7. It provides a logarithmic frequency scaled with 1 kHz and describes the log frequency in power spectrum.

$$ASC = \sqrt{\frac{\sum_{K=1}^{N/2} \log_2\left(\frac{f[k]}{1000}\right) S_r[k]}{\sum_{K=1}^{N/2} S_r[k]}} \tag{5.1}$$

ASC can be used to determine the low or high frequencies of the power spectrum.

- Audio spectrum spread (ASS): ASS is a spectral distribution at the centroid and is described by the formula as given here. It is also used to identify the difference between noise and speech.

$$ASS_r = \sqrt{\frac{\sum_{k=1}^{N/2}\left[\log_2 \frac{f[k]}{1000}\right] - ASC_r]^2 S_r[k]}{\sum_{k=1}^{N/2} S_r[k]}} \tag{5.2}$$

- Audio spectrum flatness (ASF):

ASF describes the deviation of spectral form with respect to a flat spectrum. Flat spectrum shows the noise or impulse-like signals.

- Harmonic ratio (HR): The harmonic ratio is the greatest value of autocorrelation inside the frame.

5.4   CHARACTERISTICS OF SPEECH RECOGNITION SYSTEM

Basic terminologies in speech recognition system are as follows:

- Phonemes: Basic linguistic sound unit
- Graphemes: Basic textual unit
- Utterances: An utterance is spoken word/words, a sentence, or even multiple sentences representing a single meaning. Based on the type of utterances recognized, the ASR system is classified into the following types.
  - **Isolated words:** This type of speech recognizer not only recognizes a single word but also identifies multiple words that are separated by a silent state or pause. These types of ASR system are designed with "Listen/not-Listen" conditions that require the speaker to pause between words.
  - **Connected words:** This technique is comparable to recognizing individual word but allows users to tell the word with a minimum break between them.
  - **Continuous speech:** This system allows users to talk normally, but there are several existing challenges to develop such system.
  - **Spontaneous speech:** Natural sound which happens around us that is not recorded or rehearsed is known as spontaneous speech. Examples of such systems are Alexa, Siri, and Cortana.

5.4.1   PRONUNCIATIONS

Pronunciation impacts the performance of ASR system drastically. Pronunciation is a sound of a word which is fed into a speech engine.

5.4.2 VOCABULARY

Vocabularies are dictionaries which hold list of words/utterances that are utilized by the ASR system. Smaller vocabularies are quite simple for a system to recognize, whereas large vocabularies are difficult to manage. The size or volume of the vocabulary has a fundamental outcome on the voice recognition system accuracy. Based on the number of words, vocabulary can be defined as,

- little vocabulary—10 words
- medium vocabulary—100 words
- Enormous vocabulary—1,000 words
- Extremely large vocabulary—10,000 words

5.4.3 GRAMMARS

The domain/context within which the ASR system works is defined by the grammar. The speech recognition engine utilizes a set of predefined rules, say grammar, to define the words or phrases.

5.4.4 SPEAKER DEPENDENCE

The ASR system, depending on the application, may very well be created as a speaker-reliant or autonomous framework. The speaker subordinate framework is the one that remembers one specific client voice, though the speaker-free framework fosters a recognizer which distinguishes the discourse of any client.

5.5 THE WORKING OF A SPEECH RECOGNITION SYSTEM

Figure 5.2 depicts the speech recognition process.

5.5.1 INPUT SPEECH

The analog signal captured using microphone can be digitized using sound card. A spectrogram is a time-based depiction of a voice signal. The flat pivot of a spectrogram shows time; the upward hub portrays the recurrence or power of the information-expressed stream. In the creation of voice to text models, it is a widely utilized representation of voice signal. The following Figure 5.3 depicts a time–frequency representation of a spoken signal.

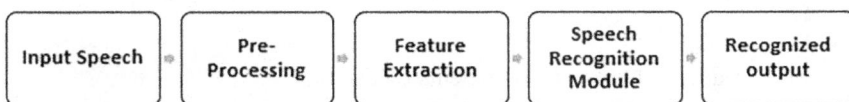

| Input Speech | → | Pre-Processing | → | Feature Extraction | → | Speech Recognition Module | → | Recognized output |

FIGURE 5.2 Speech Recognition System.

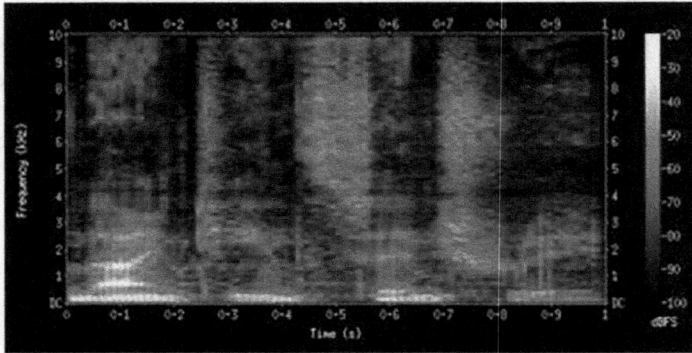

FIGURE 5.3   Spectrogram Representation.

5.5.2   AUDIO PRE-PROCESSING

Audio pre-processing involves the following steps:

- Denoising or noise reduction: Noise removal involves the reduction of background ambient noise or noise at the time of recording.
- Audio signal framing: Splitting of input audio into small frames or blocks.
- Windowing: The input frames are chunked into small signals, usually window size of 20–25 ms, sliding at a rate of 10 ms.
- Normalization: Used to minimize speaker and recording variability while maintaining the features' discriminative strength.

5.5.3   FEATURE EXTRACTION

Feature extraction is utilized to extract a set of acoustic features in speech signal. Such features can be computed by processing the speech signal waveform. The two variants of features are prosodic feature and spectral features which can be extracted from an input speech. Prosodic features are the aspects of each signal, which deals with auditory qualities of sound. Spectral features are frequency-based features.

5.6   AUDIO FEATURE EXTRACTION TECHNIQUES

Extracting the features after pre-processing, the input signal minimizes the quantity of data for analysis. Widely used speech feature extraction techniques are:

- Spectrogram
- Mel Frequency Cepstral Coefficients (MFCC)
- Short-Time Fourier Transform

5.6.1 SPECTROGRAM

A spectrogram is a graphical illustration of the amplitude of a sound that graphs the signal's constituent frequencies versus time or some other variable. Spectrograms are essentially two-dimensional graphs with color representing a third dimension. Two-dimensional signals are created from the spoken signal. In a spector graph, time is displayed along horizontal axis from left to right. The lowest frequency will be at the bottom, and the highest frequency at the top of the vertical axis, which symbolizes frequency that can also be thought of as pitch or tone. The workflow of obtaining spectrogram is depicted in Figure 5.4, and Python code for spectrogram is given here followed by the output in Figure 5.5.

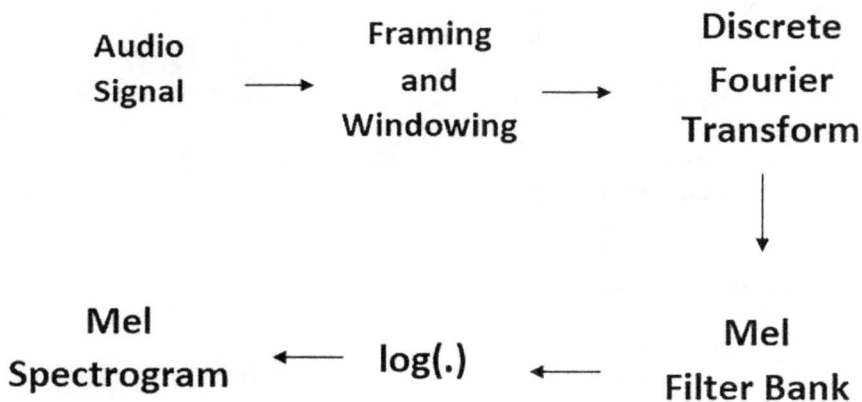

FIGURE 5.4   Spectrogram Pipeline.

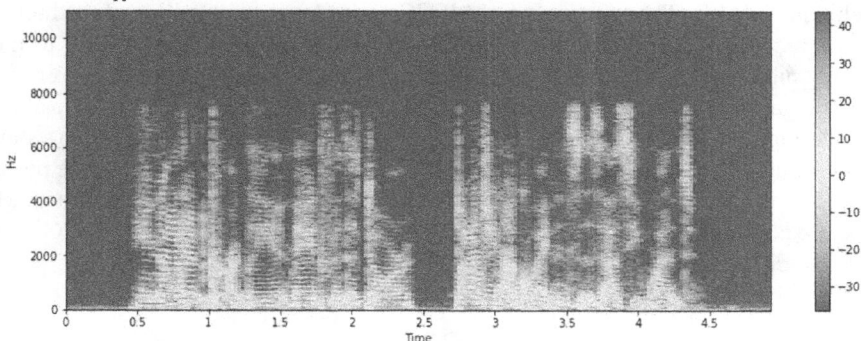

FIGURE 5.5   Spectrogram Output.

Python code for spectrogram extraction is as follows:

```
Importing Libraries7
import librosa
import glob
%matplotlib inline
import matplotlib.pyplot as plot
import librosa.display
import IPython.display as ipdis
Loading an audio
for filename in glob.glob('/content/drive/MyDrive/LibriSpeech/dev-clean/1462/*/*.
    wav'):
speech_path = filename
x, sr = librosa.load(speech_path)
Plotting an audio in waveform
plot.figure()
librosa.display.waveplot(x, sr=sr)
time.sleep(0.1)
plot.pause(0.0001)
Plotting an audio in Spectrogram

X = librosa.stft(x)
Xdb = librosa.amplitude_to_db(abs(X))
print(Xdb)
plot.figure(figsize=(14, 5))
librosa.display.specshow(Xdb, sr=sr, x_axis='time', y_axis='hz')
plot.colorbar()
```

5.6.2   MFCC

The Mel Frequency Cepstral Coefficients (MFCCs) is a well-known feature extraction technique in ASR. The functioning of MFCC is similar to the working of the human ear. It records every tone with a real frequency f (Hz), which corresponds to mel scale's subjective pitch. The MFCC gives a discrete cosine change of the energy sign's logarithm on a mel recurrence scale. At first, the discourse signal is separated into casings, and, afterward, Fast Fourier change (FFT) is used to each edge to secure power range. The Mel scale then, at that point, applies the channel bank to the power range. In the wake of switching the power range over completely to log space, the discrete cosine transform is applied to the discourse sign to acquire the MFCC coefficients. Eq. (5.3) is used to resolve the mel function for any recurrence.

$$mel(f) = 2595 \, x \log_{10}\left(1 + \frac{f}{700}\right) \tag{5.3}$$

The MFCC coefficients are determined by Eq. (5.4).

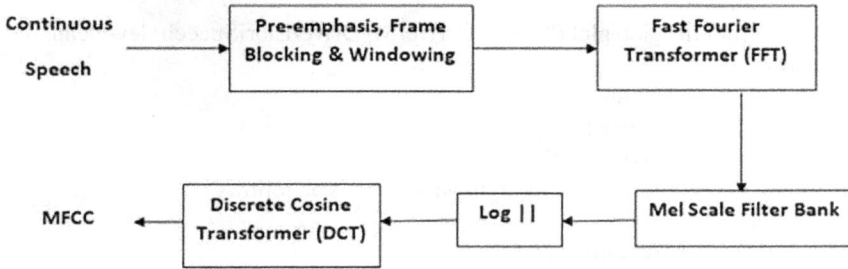

FIGURE 5.6 MFCC Pipeline.

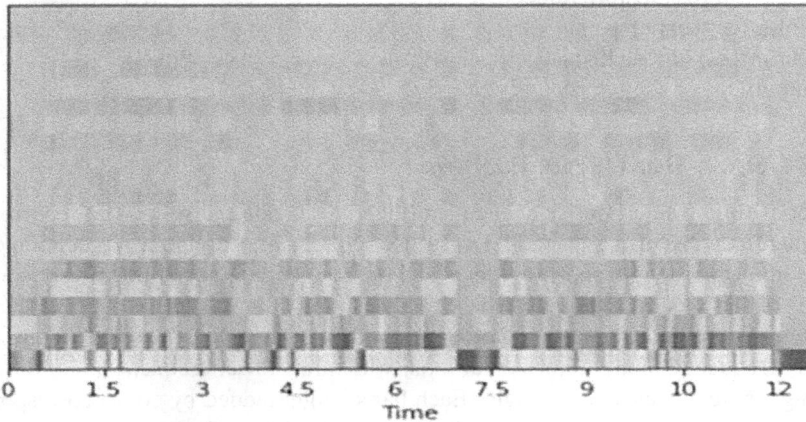

FIGURE 5.7 MFCC Representation.

$$\widehat{C}_n = \sum_{n=1}^{k} (\log \widehat{s}_k) \cos\left[n\left(k - \frac{1}{2}\right)\frac{\dot{A}}{k}\right] \qquad (5.4)$$

Figure 5.6 is the workflow diagram of MFCC. It defines all the steps to obtain the MFCC coefficients. For audio signals with background sound, MFCC function does not work effectively and so is not suited for such powerful speech recognition system. Python code for extracting MFCC is given here followed by the output in Figure 5.7.

Python code for MFCC extraction for importing libraries is as follows:
```
import librosa
import glob
%matplotlib inline
import matplotlib.pyplot as plt
import librosa.display
import IPython.display as ipd
```

Loading an audio:

```
for filename in glob.glob('/content/drive/MyDrive/LibriSpeech/dev-clean/*/*/*.
    wav'):
speech_path = filename
x, sr = librosa.load(speech_path)
print(type(x), type(sr))
```

Python code for plotting an audio in waveform is as follows:

```
plt.figure()
librosa.display.waveplot(x, sr=sr)
time.sleep(0.1)
plt.pause(0.0001)
```

MFCC coefficient:

```
mfccs = librosa.feature.mfcc(x, sr=sr)
print(mfccs.shape)
print(mfccs) librosa.display.specshow(mfccs, sr=sr, x_axis='time')
```

Output:

5.6.3 SHORT-TIME FOURIER TRANSFORM

The brief time frame Fourier change is utilized to decide the sinusoidal recurrence and stage content of a sign (STFT). To use STFTs, first divide a larger temporal signal into equal-length short segments and then apply the Fourier transform to each of the smaller pieces separately. Music analysis typically makes use of STFTs, conventional Fourier transforms, and other methods. For instance, the spectrogram can display frequency along a horizontal axis, which has lower frequencies on the left and higher frequencies on the right. Each bar's height (added by color) corresponds to the amplitude of the frequencies falling inside that band. Each new bar represents a unique morph, and the depth dimension indicates time. This type of visual aid is used by audio engineers to learn more about an audio sample, such as the frequencies of particular noises or frequencies that could be more or less resonant in the environment where the signal was captured. The pipeline for STFT is in Figure 5.8 and its representation is shown in Figures 5.9 and 5.10.

```
Python code:
import librosa
speech_path = '/content/demo.wav'
x, sr = librosa.load(speech_path)
print(type(x), type(sr))
librosa.load(speech_path, sr=44100)
import IPython.display as ipd
ipd.Audio(speech_path)
%matplotlib inline
import matplotlib.pyplot as plt
import librosa.display
plt.figure(figsize=(14, 5))
librosa.display.waveplot(x, sr=sr)
```

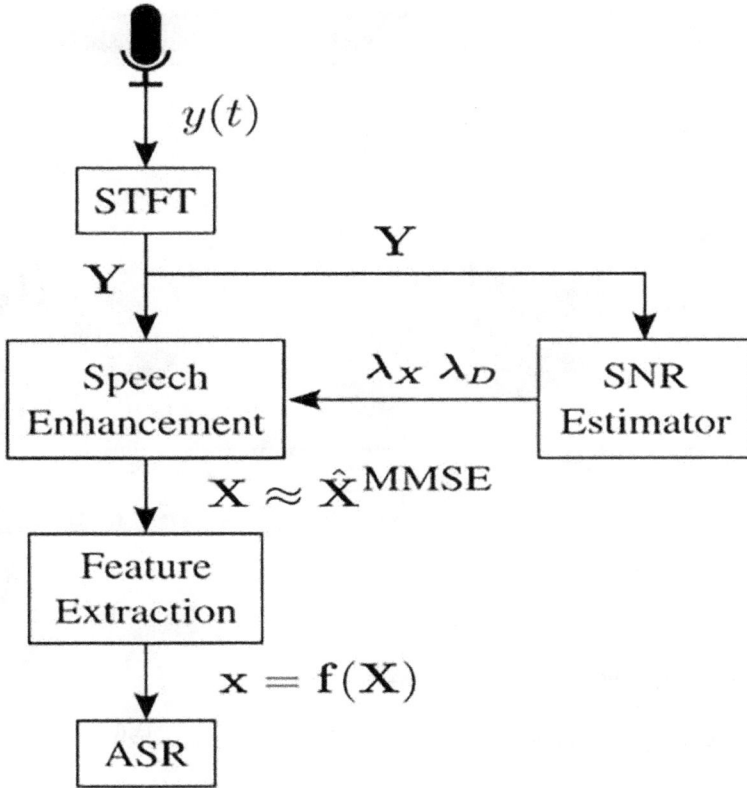

FIGURE 5.8 Fourier Transform.

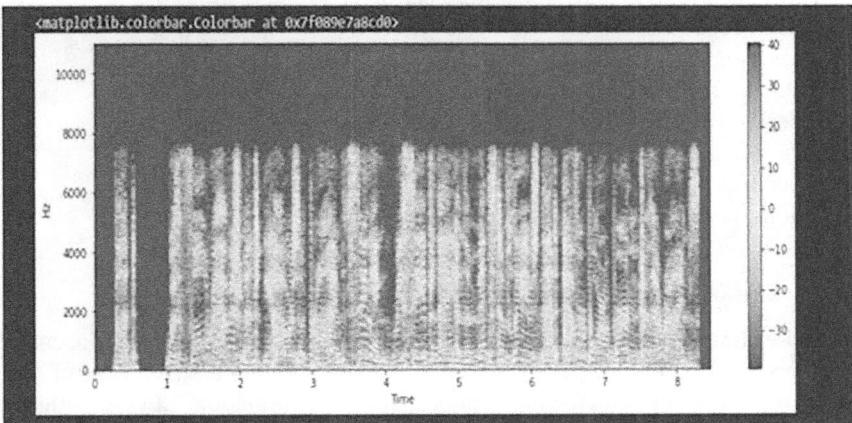

FIGURE 5.9 STFT Representation.

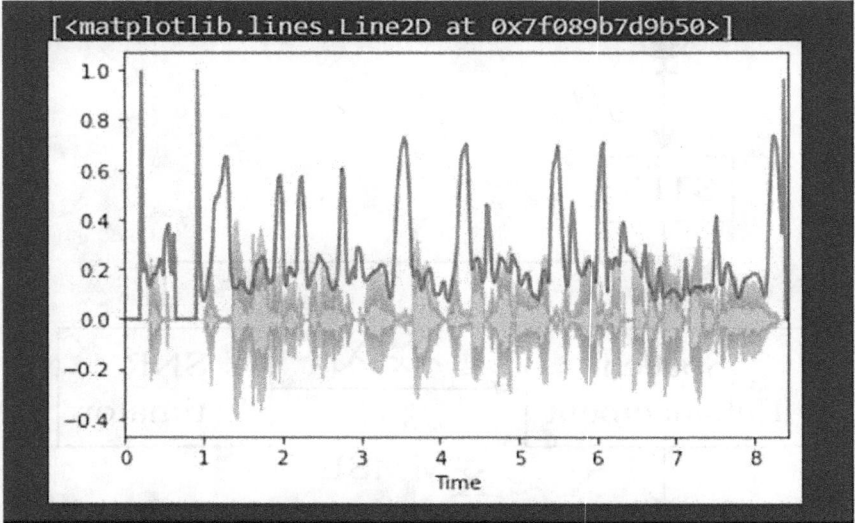

FIGURE 5.10   STFT Feature Representation.

```
X = librosa.stft(x)
Xdb = librosa.amplitude_to_db(abs(X))
plt.figure(figsize=(14, 5))
librosa.display.specshow(Xdb, sr=sr, x_axis='time', y_axis='hz')
plt.colorbar()
x, sr = librosa.load(speech_path)
plt.figure(figsize=(14, 5))
librosa.display.waveplot(x, sr=sr)
import sklearn
spectral_centroids = librosa.feature.spectral_centroid(x, sr=sr)[0]
spectral_centroids.shape
frames = range(len(spectral_centroids))
t = librosa.frames_to_time(frames)
def normalize(x, axis=0):
return sklearn.preprocessing.minmax_scale(x, axis=axis)
librosa.display.waveplot(x, sr=sr, alpha=0.4)
plt.plot(t, normalize(spectral_centroids), color='r')
```

5.6.4   LINEAR PREDICTION COEFFICIENTS (LPCC)

Cepstral examination is typically applied for discourse handling since it accurately represents the discourse waveform. As displayed in Figure 5.11, the means of LPCC are outline hindering, windowing, autocorrelation examination, LPC investigation, and LPC boundary change.

LPCC is calculated using Eq. (5.5),

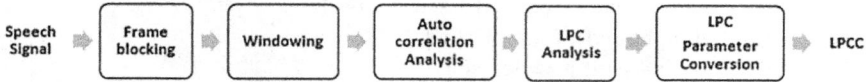

FIGURE 5.11   LPCC Processor.

$$C_m = a_m + \sum_{k=1}^{m-1} \left[\frac{k}{m} \right] c_k \, a_{m-k} \qquad (5.5)$$

The cepstral coefficient is C_m, and the linear prediction coefficient is. Because LPCC features are less susceptible to noise than MFCC features, they have a lower word error rate. Python code for LPCC is as follows:

```
Importing libraries:
import numpy as np
from scipy.io.wavfile import read
import matplotlib.pyplot as plt
import glob
import soundfile as sf
Functions to find LPCC:
def autocorr(self, order=None):
if order is None:
order = len(self) — 1
autocor=[]
sum=0
for tau in range(order+1):
for n in range(len(self)—tau):
sum +=self[n] * self[n + tau]
autocor.append(sum)
return autocor
def lpc(seq, order=None):
# In this lpc method we use the slow(if the order is >50) autocorrelation approach.
acseq = np.array(autocorr(seq, order))
# Using pseudoinverse to obtain a stable estimate of the toeplitz matrix
a_coef = np.dot(np.linalg.pinv(scipy.linalg.toeplitz(acseq[:-1])), -acseq[1:].T)
# Squared prediction error, defined as e[n] = a[n] + \sum_k=1^order (a_k * s_{n-k})
err_term = acseq[0] + sum(a * c for a, c in zip(acseq[1:], a_coef))
return a_coef.tolist(), np.sqrt(abs(err_term))
def lpcc(seq, err_term, order=None):
if order is None:
order = len(seq) — 1
lpcc_coeffs = [np.log(err_term), -seq[0]]
for n in range(2, order + 1):
# Use order + 1 as upper bound for the last iteration
upbound = (order + 1 if n > order else n)
```

```
lpcc_coef = -sum(i * lpcc_coeffs[i] * seq[n—i—1]
for i in range(1, upbound)) * 1./upbound
lpcc_coef -= seq[n—1] if n <= len(seq) else 0
lpcc_coeffs.append(lpcc_coef
return lpcc_coeffs
Computing LPCC values:
order = 12
for count,filename in enumerate(glob.glob('/content/drive/MyDrive/LibriSpeech/
    dev-clean/1272/128104/*.wav')):
sr,wav = read(filename)
lpc_value,err = lpc(wav,order)
lpcc_value = lpcc(lpc_value,err,order)
print(lpcc_value)
plt.figure(count)
plt.xlabel("Time")
plt.ylabel("LPCC Coeff")
plt.plot(lpcc_value)
plt.show()
```

5.6.5　Discrete Wavelet Transform (DWT)

The wavelet transform is a popular signal-processing method in the frequency and time domains. Equations (5.6) and (5.7) define the wavelet function and scaling function of the DWT. The variants of DWT parameter is depicted in Figure 5.12.

$$\varphi(t) = \sum_{n=0}^{N-1} h[n]\sqrt{2}\varphi(2t-n) \tag{5.6}$$

$$\rho(t) = \sum_{n=0}^{N-1} g[n]\sqrt{2}\varphi(2t-n) \tag{5.7}$$

where $h[n]$ is the low-pass filter's impulse response, $g[n]$ is the high-pass filter's impulse response, and $\varphi(t)$ is the scaling function and $\rho(t)$ is the wavelet function, respectively.

The DWT for a continuous signal is given in Eq. (5.8).

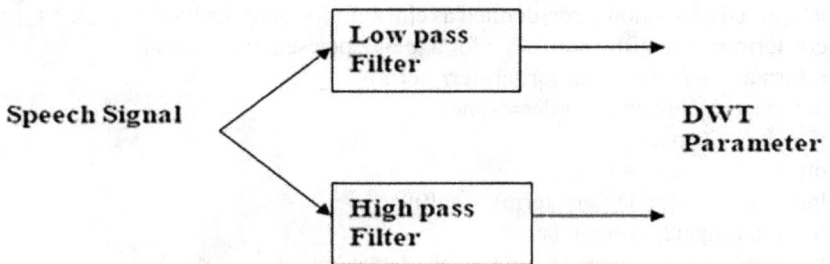

FIGURE 5.12　Discrete Wavelet Transform.

$$(DWT)(m,p) = \int_{-\infty}^{+\infty} x(t).\varphi_{m,p} dt \qquad (5.8)$$

where $\varphi_{m,p}$ is the wavelet function bases, m is the dilation parameter, and p is the translation parameter. $\varphi_{m,p}$ in Eq. (5.9) is

$$\varphi_{m,p} = \frac{1}{\sqrt{a_0^m}} \varphi\left(\frac{t - pb_0 a_0^m}{a_0^m}\right) \qquad (5.9)$$

The DWT of a discrete sign is in condition (5.10).

$$(DWT)(m,k) = \frac{1}{\sqrt{a_0^m}} \sum_n x[n].g\left(\frac{n - nb_0 a_0^m}{a_0^m}\right) \qquad (5.10)$$

where g (*) is the mother wavelet and $x[n]$ is the discretized signal.

5.6.6 Perceptual Linear Prediction (PLP)

PLP's starting points can be followed to the nonlinear bark scale, which is expected to further develop discourse acknowledgment by eliminating speaker-subordinate components. PLP produces a smooth momentary range which is level and compact as in human hearing. PLP is characterized by direct expectations for otherworldly smoothing, also known as perceptual straight expectation. It combines linear prediction analysis and spectral analysis. Figure 5.13 shows the steps involved in computing the PLP features.

Table 5.1 presents a comparative analysis of six feature extraction techniques. It provides characteristics of the different feature extraction techniques which enable users to select the appropriate technique depending on the scenario. The characteristics include computational speed, noise resistance, sensitivity, and reliability.

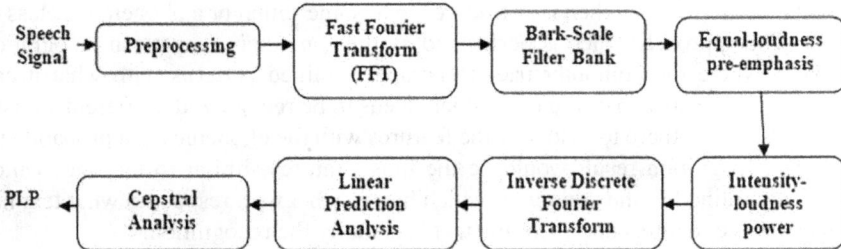

FIGURE 5.13 PLP Processor.

TABLE 5.1

Comparison of Six Feature Extraction Techniques

	Type of Filter	Shape of Filter	Speed of Computation	Type of Coefficient	Noise Resistance	Reliability
MFCC	Mel	Triangular	High	Cepstral	Medium	High
LPC	Linear prediction	Linear	High	Auto correlation Coefficient	High	High
LPCC	Linear prediction	Linear	Medium	Cepstral	High	Medium
LSF	Linear prediction	Linear	Medium	Spectral	High	Medium
DWT	Low pass and High pass	–	High	Wavelets	Medium	Medium
PLP	Bark	Trapezoidal	Medium	Cepstral & Auto Correlation	Medium	Medium

By removing relevant information from voice signals, these feature extraction techniques enable speech recognition. Speech signals with short durations between 5 and 100 ms are considered to be of sufficient length. The feature extraction techniques like MFCC, LPCC, and PLP are typically employed for short-time spectral analysis. The noise is a significant obstacle for voice recognition. For noisy robust ASR models, techniques like STFT and log mel filter bank energies are adopted. Hence, selecting the appropriate feature extraction technique greatly varies the performance of ASR models.

5.7   STATISTICAL SPEECH RECOGNITION

The computer first transcribes speech and tries to make out sense of what has been transcribed. The background knowledge could span over the speech production process. Human knowledge has to be given to a machine in order to automatically recognize speech. The speech signals are first transformed into features by means of a feature extraction algorithm. These features are usually calculated for time span of approximately 20 to 30 ms. That is, we calculate one vector of features for each 20 or 30 ms. These features then make reference to some sound or a phoneme, a class of sounds. The approach which is performed by the computer is a pattern recognition approach, where the computer tries to compare trained patterns with what it can observe from the speech of the user which needs to be recognized. Different statistical algorithms are there to associate the features with the phonemes in a probabilistic way. The recognition result would be the most features similar to the ones which are already trained in the computer, which have a label with respect to what text has produced these features and then this text is actually the recognition.

Figure 5.14 represents the fundamental components of a speech recognition system. The speech recognizer needs to be trained and needs to store which types of

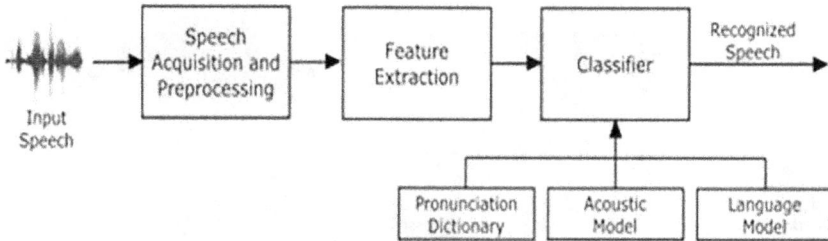

FIGURE 5.14   Fundamental Components of Speech Recognition.

features are representative for certain phonemes, and this is called the acoustic model of the speech recognizer. The sequence of probable phonemes needs then to be transformed into a sequence of words or even a sentence which is to be recognized. This is done in the decoding step. In the decoder, the phoneme probabilities are organized in order to form words and sentences. So the decoder needs to access a vocabulary, but it needs to know which words can follow each other. A language model/grammar is employed in order to organize the words in a certain sequence. To extract the features for ASR processes, first we try to separate the excitation signal and the vocal track modulation from the speech and then we make use of the hearing characteristics of the human ear in order to process in a way which is similar to what we have in the human ear. The next step is to classify the features into phonemes and then to organize these phonemes into words or sequence of words. The two popular approaches which are used for this task are Hidden Markov Model and Neural Network. Both of them are statistical approaches that calculate a probability that a certain feature vector is related to certain phoneme. So it gives probabilities of phonemes, and these probabilities of phonemes can then be organized into probabilities of words. The input to an ASR system is the raw one-dimensional speech signal. The fundamental unit of ASR is phonemes or phones. Word models can be built by concatenating phone models. Let x represent an input audio sample and the function f(x) that maps the sequence of words to the transcripts of the speech signal. A basic speech recognition system includes following units: pre-processing, feature extraction, acoustic modelling, and language modelling for ASR as shown in Figure 5.15. Acoustic model converts the speech into their corresponding phonemes. The lexicon or pronunciation model converts the phones to corresponding words. Language model defines the most likely sequence of words.

5.7.1   Acoustic Model

It is classifier that tries to identify the pronunciation of speech spoken, on the basis of input feature. Raw audio waveform of speech is feed as input. Acoustic modelling involves an audio signal that is segmented into smaller time frames that is of around 25 ms. Each frame is examined by acoustic models, which provide the possibility of various phonemes that can be used in that particular audio portion. Acoustic models are necessary since there are numerous ways that people can

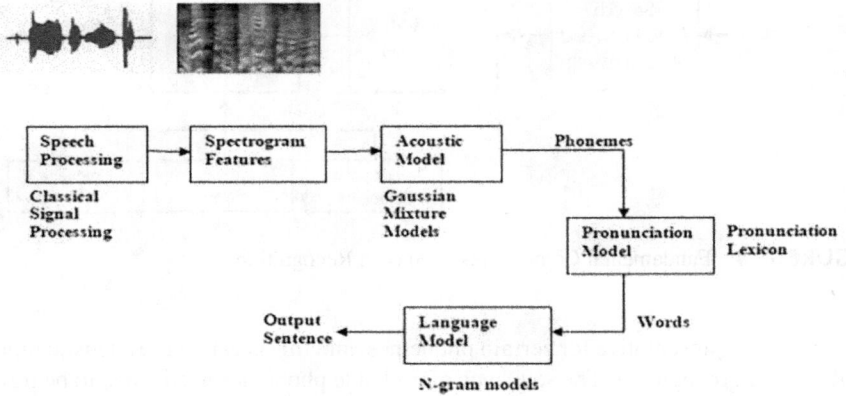

FIGURE 5.15 Architecture of ASR System.

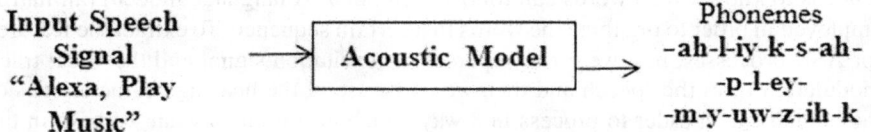

FIGURE 5.16 Acoustic Modeling.

pronounce the same word. The same speech can sound different on the basis of the background of the speaker due to elements like background sound and language. To establish the relationship between audio frames and phonemes, acoustic models employ deep-learning algorithms that have been trained on hours of various audio recordings and relevant texts. Figure 5.16 depicts the working of acoustic model with an example.

Assuming the acoustic feature vectors of a speech signal, $\hat{x} = \{x_1, x_2, \ldots x_n\}$ the idea of ASR is to identify the word sequence $\hat{w} = \{w_1, w_2, \ldots w_n\}$ $\hat{w}$. is defined as

$$\hat{w} = argmax \, P(W \mid X) \tag{5.11}$$

where the probability of word given x is given as $P(W \mid X)$. According to Bayes rule, condition (5.2) can be revamped as

$$\hat{w} = argmax \frac{P(X \mid W) P(W)}{P(X)} \tag{5.12}$$

where $P(X \mid W)$ represents the distribution of speech features and acoustic vectors in a speech signal, also known as the acoustic model. $P(W)$ is the prior probability of the text sequence, and, in context to ASR system, it denotes the language model. By applying this to ASR, the statistical ASR system is defined as in Eq. (5.4):

$$P(W|X) = \frac{P(X|W)P(W)}{P(X)}$$

(5.13)

Substituting Eq. (5.4) in Eq. (2),

$$\hat{w} = argmax\, P(X|W).P(W)$$

(5.14)

The acoustic models identify the phoneme sequence, given the feature vector.

$$\hat{w} = \underset{W}{argmax}\, P(X|W)P(W)$$

$$= \underset{W \in V^*}{argmax} \sum_{S} P(X,S|W)P(W)$$

(5.15)

$$\approx \underset{W,S}{argmax}\, P(X,S)P(S|W)P(W)$$

Eq. (5.15) shows the acoustic and language model where $P(X,S)$ represents the acoustic function of each phone state, $P(S|W)$ represent the earlier likelihood of a word.

5.7.2 PRONUNCIATION MODEL

Pronunciation lexicon converts the phoneme sequence to its corresponding words as shown in Table 5.2. By isolating the sound bite with a sliding window, a grouping of sound edges is produced. Figure 5.17 shows the pronunciation model with an example by converting phoneme sequence to words.

The probabilistic chain rule is used in lexicon model as shown in Eq. (5.16), where W is the words and S is the phoneme sequence.

$$P(S|W) = \prod_{t=1}^{T} P(s_t | s_{t-1}, W)$$

(5.16)

TABLE 5.2
Lexicon or Dictionary

Phone Sequence	Words
l-ay-k	Like
g-uh-d	good
Ih-z	is
f-ay-v	Five

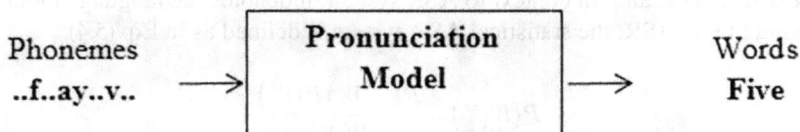

FIGURE 5.17   Lexicon or Pronunciation Model.

5.7.3 Language Model

In language demonstrating, various measurable and probabilistic methods are utilized to compute the probability that a given series of words will show up in an expression. The noise is speech recognition. The language model predicts the likelihood of each word in a phrase on the basis of the output of the pronunciation model and then converts the words into sentences. Language model helps to improve the accuracy of ASR system. It is represented as a probability distribution $P(W)$ which reflects the frequency a string w occurs as a sequence. $P(W)$ can be represented as

$$
\begin{aligned}
P(W) &= \{w_1, w_2, \ldots w_n\} \\
&= P(w_1) P(w_2|w_1) P(w_3|w_1 w_2) P(w_n \mid w_1 w_2 \ldots w_n)\} \\
&= \prod_{i=1}^{n} P(w_i|w_1 w_2 \ldots w_{i-1})
\end{aligned}
\tag{5.17}
$$

where $P(w_i|w_1 w_2 \ldots w_{i-1})$ is the probability *of* w_i, given $w_1 w_2 \ldots w_{i-1}$. In other words, the word w_i depends on the past history or previous words.

It is used to correct the spelling errors in ASR system by identifying the higher probable words as shown here.

Language Model in ASR Spell Correction:

$$
P(Speech\ \textbf{\textit{Recogntion}}\ System) < P(Speech\ \textbf{\textit{Recognition}}\ System)
\tag{5.18}
$$

The following are some of the variants of language models:

- N-gram: Unigram and bigram are the variants of n-gram. For instance, given bigram of prior words, it will predict the next most likely word.

$$
Uni - Gram = P(w_3|w_2) = p(System \mid Recognition)
\tag{5.19}
$$

$$
Bi - Gram = P(w_3|w_1\ w_2) = p(System \mid Speech\ Recognition)
\tag{5.20}
$$

$$
n - Gram = P(w_n|w_3\ w_1) = p(System \mid Speech\ Recognition)
\tag{5.21}
$$

ASR uses n-gram language models to guide the speech for the correct word sequence. It predicts the likelihood of the nth word using the previously occurring words. Commonly used n-gram models are trigrams where n is 3, and it is represented as $P(w_3 \mid w_1, w_2)$, Bigram model is represented as $P(w_2 \mid w_1)$. To estimate $P(w_i \mid w_{i-1})$ (i.e.) the probability of the word w_i, given w_{i-1}, simply count the occurrence of the sequence $P(w_i \mid w_{i-1})$ and then normalize the count by the number of times w_{i-1} occurs. In the trigram model, the likelihood of a word is determined by the two words preceding it. For example,

$$p(w) = p(w_1)p(w_2|w_1)p(w_3|w_2w_1) \tag{5.22}$$

The probabilities for trigram model are computed by the frequencies of the word pair $c(w_{i-2}, w_{i-1})$.

$$p(w_i|w_{i-2}w_{i-1}) = \frac{c(w_{i-2}, w_{i-1}, w_i)}{c(w_{i-2}, w_{i-1})}. \tag{5.23}$$

N-gram model makes use of three types of decoding search to select the most probable candidate such as:

- Beam search
- Greedy search
- Neural rescoring: Neural-network-based language model which utilizes a deep learning-based approach to identify the most likely sequence.
- Transformer-based BERT: Hugging face library provides transformer-based BERT model which is an extensively used in ASR system. It is used to grade the delivered sentence of the speech.

5.7.4 CONVENTIONAL ASR APPROACHES

- *Acoustic–phonetic approach*: The acoustic–phonetic method begins with a phantom investigation of the discourse, followed by highly recognizable proof, which transforms the unearthly information into a bunch of elements that characterize the overall acoustic properties of the individual phonetic units. The division and naming stage that follows separates the discourse signal into stable acoustic locales, delivering a phoneme cross-section portrayal of the discourse by doling out at least one phonetic mark to each divided district. The strategy's last stage endeavors to find a significant word utilizing the phonetic mark groupings created by division and naming (or series of words).
- *Generative learning approach—HMM-GMM*: The GMM-based hidden Markov models (HMMs) are the foundation of the generative learning approach. HMMs based on Gaussian mixture models (GMMs) capture the sequential structure of speech signals in traditional speech

recognition systems. Chapter 6 discusses the conventional ASR models in detail.

- *Deep-learning-based approach*: Neural networks such as RNN, Long Short Term Memory (LSTM), and transformer-based techniques are widely utilized for developing end-to-end ASR systems. Chapter 7 discusses the deep learning-based cutting-edge ASR models in detail.

5.8 SPEECH RECOGNITION APPLICATIONS

As a result of the massive increase in cloud computing, data, and processing power, speech analytics has advanced to the point where many difficult situations are becoming a reality. Voice control in home entertainment systems and speech-centric information processing tools, such as Siri on the iPhone, Bing voice search on Windows Phones, and Google Now on Android, are examples (such as Kinect on the Xbox). Dictation systems, speech user interfaces, voice dialing, call routing, home appliance management, command and control, voice-assisted search, simple data entry, hands-free applications, and learning systems for people with disabilities are examples of typical applications, as shown in Figure 5.18.

The major applications using the speech recognition technology are as follows:

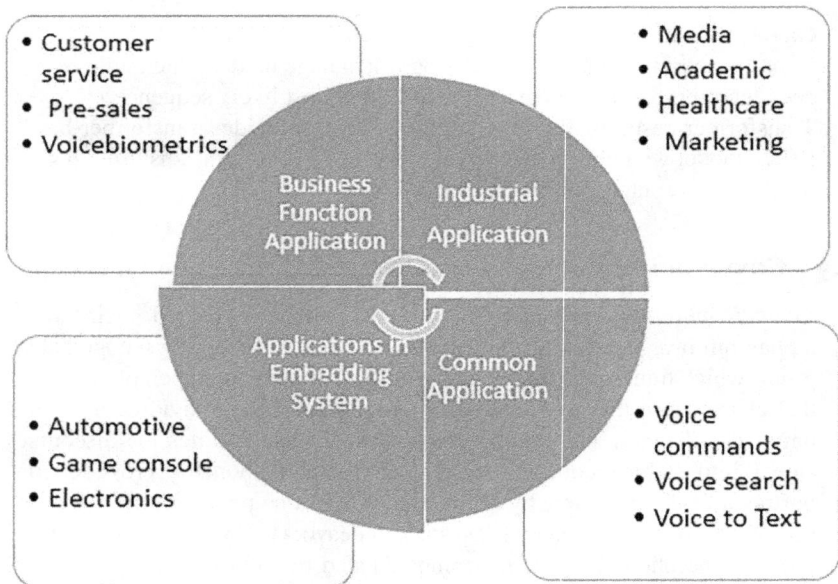

FIGURE 5.18 Applications of ASR.

5.8.1 IN BANKING

- In the banking, voice-enacted banking could reduce the need for human client assistance and reduce labor expenses.

5.8.2 IN-CAR SYSTEMS

- Straightforward voice orders can be utilized to answer calls, change radio broadcasts, and play music from a MP3 player, streak drive, or viable cell phone. The ability to recognize voice varies by car model and make.

5.8.3 HEALTH CARE

- Discourse acknowledgment programming can be beneficial to people with disabilities. Discourse acknowledgment programming is used to automatically make closed inscriptions for exchanges, for example, gathering room conversations, school addresses, and strict administrations for people who are deaf or hard of hearing.
- Medical documentation: Discourse acknowledgment can be coordinated into the front-end or back-end of the clinical documentation process in the medical services industry.

5.8.4 EXPERIMENTS BY DIFFERENT SPEECH GROUPS FOR LARGE-VOCABULARY SPEECH RECOGNITION

- Bing-Voice-Search speech recognition task

Bing-Voice-Search, the Bing mobile voice search application, was used to collect data for the first successful DBN-DNN and DNN-HMM acoustic models for a large vocabulary speech recognition challenge (BMVS).

- Switchboard speech recognition task

Switchboard is a transparently accessible benchmark task for discourse-to-message recording that takes into consideration substantially more exhaustive examinations of strategies. The DNN-HMM acoustic model, which had recently been prepared with as much as 48 hours of information and many tied triphone states as targets, has now been prepared with well over 300 hours of information and a large number of tied triphone states as targets.

- Google voice input speech recognition task

The Google voice input discourse acknowledgment task deciphers cell phone client exercises like short messages, messages, and voice search demands. Given the size of the jargon in question, a language model fit for taking care of both transcription and search questions is being utilized. The acoustic models are triphone frameworks created from choice trees that utilize GMMs with differing quantities of Gaussians per acoustic state. It utilizes a three-state, left-to-right GMM-HMM with setting subordinate cross-word triphone HMMs.

- YouTube speech recognition task

The YouTube speech recognition challenge's goal is to translate YouTube content. A total of 1,400 hours of aligned training data were generated using Google's full-fledged baseline, which was built with a much larger training set. This was used to build a new baseline system using nine frames of MFCCs that had undergone LDA transformation. Triphone states numbering 17, 552 were discovered using decision-tree clustering and speaker adaptive training.

5.8.5 MEASURE OF PERFORMANCE

The effectiveness of ASR is identified in the perspective of accuracy and latency. Accuracy is identified using word error rate (WER), character error rate (CER), and word recognition rate (WRR), which are calculated using Eqs. (5.19) and (5.20). Latency is used to measure the performance of streaming ASR system.

$$Word\ Error\ Rate(\%) = \frac{Insertion(I) + Substitution(S) + Deletion(D)}{No\ of\ Reference\ Words(N)} *100 \quad (5.19)$$

$$Word\ Recognition\ Rate = 1 - WER = \frac{N - S - D - I}{N} \quad (5.20)$$

5.9 CHALLENGES IN SPEECH RECOGNITION

Robustness refers to a speech recognition system's ability to effectively manage various types of fluctuation in the voice stream. The system can achieve high performance in limited circumstances. Under uncontrolled conditions, it is difficult to achieve higher accuracy. The performance of the system can differ according to the following conditions explained in the next sections.

5.9.1 VOCABULARY SIZE

It is rather simple to distinguish a term from a tiny set of vocabulary. Small vocabulary speech recognition system (usually less than 100 words) is used in command and control applications such as Interactive Voice Response (IVR), voice dialing, and providing instructions. Large vocabulary speech recognition system (usually more than 5,000 words) is used for continuous speech recognition, captioning live audio/video programs and so on.

5.9.2 SPEAKER-DEPENDENT OR -INDEPENDENT

It is always interesting to study the response of the system when some unknown speaker evaluates it. The system is trained with some parameters that are highly speaker specific. Usually it is difficult to attain same performance with Speaker-Independent (SI) system as compared to Speaker-Dependent (SD) system.

5.9.3 ISOLATED, DISCONTINUOUS, AND CONTINUOUS SPEECH

Initially, single word recognition are conducted on isolated voice data. Full sentences in which words are purposefully divided by silence are referred to as discontinuous speech. Continuous speech refers to sentences that are naturally spoken. Because word boundaries appear to be spotted and words are nicely articulated, isolated and discontinuous speech recognition is a relatively easier process. However, in continuous speech, word boundaries are hazy, and co articulation affects pronunciation. When the system meets spontaneous speech such as coughing, emotions, erroneous starts, "um", "hm," as an incomplete sentence, and so on, the work becomes more difficult.

5.9.4 PHONETICS

Phoneme, in linguistic, is the fundamental unit of speech that is combined with other phonemes to form a meaningful word. Phonemes are language dependent, and an ASR system should have proper phonetic knowledge about the particular language.

5.9.5 ADVERSE CONDITIONS

Usually, a system is trained with clean acoustic data. A number of unfavorable circumstances have a significant impact on the system's operation. These are background noise, acoustic mismatch, channel noise, and so on.

5.10 OPEN-SOURCE TOOLKITS FOR SPEECH RECOGNITION

For the purpose of developing a recognition system, researchers studying ASR have a variety of open-source toolkit options. The openly available toolkits for ASR system listed here are some of the more popular.

5.10.1 FRAMEWORKS

- Sphinx: Carnegie Mellon University developed an ASR toolbox.
- Kaldi: A free and open-source C++ ASR framework designed for both academic and commercial speech processing.
- NVIDIA's Nemo Kit, Jasper, and Quartz ASR pre-trained models are available at NVIDIA NGC portal.

- ESPnet: Inspired by Kaldi, an end-to-end deep-learning-based ASR system is created with PyTorch and Chainer backends.
- HTK, Julius (both written in C).

5.10.2 Additional Tools and Libraries

- KenLM: A set of high-performance tools for modelling n-gram languages, typically used in conjunction with ASR frameworks.
- LIME: Local interpretable model-agnostic explanation is a machine learning and deep learning explainer that is both local and model-agnostic.
- SoX: A library with tools for audio manipulation. It supports a variety of file types and may be used to play, convert, and manipulate audio files.
- LibROSA: A well-known Python audio analysis module that is used for digital signal processing and feature extraction (DSP).

5.11 SUMMARY

This chapter covers the speech recognition basics, characteristics of ASR, types of ASR, and milestones of various developments in ASR. Classification of the audio signals is based on some audio parameters which are the audio features. This chapter also narrates the complete framework of speech recognition system which includes speech capturing, pre-processing, different audio features, various feature-extraction techniques, and models of speech recognition systems. It highlights on the various applications of speech recognition and the benchmark datasets for speech recognition together with the exposure on various open-source toolkits for speech recognition.

BIBLIOGRAPHY

Afouras, T., Chung, J.S., Senior, A., Vinyals, O., and Zisserman, A. 2018. Deep audio-visual speech recognition. *IEEE Transactions on Pattern Analysis and Machine Intelligence*, *44*, pp. 8717–8727.

Feng, W., Guan, N., Li, Y., Zhang, X., and Luo, Z. 2017, May. Audio visual speech recognition with multimodal recurrent neural networks. In *2017 International Joint Conference on Neural Networks (IJCNN)* (pp. 681–688). New York: IEEE.

Kerkeni, L., Serrestou, Y., Mbarki, M., Raoof, K., Mahjoub, M.A., and Cleder, C. Automatic speech emotion recognition using machine learning. In *Social Media and Machine Learning*. IntechOpen, Edited by Alberto Cano, 2019. https://doi.org/10.5772/intechopen.84856.

Makino, T., Liao, H., Assael, Y., Shillingford, B., Garcia, B., Braga, O., and Siohan, O. 2019. Recurrent neural network transducer for audio-visual speech recognition. In *2019 IEEE Automatic Speech Recognition and Understanding Workshop (ASRU)* (pp. 905–912). New York: IEEE.

Miao, Y., and Metze, F. 2016. Open-domain audio-visual speech recognition: A deep learning approach. In *Interspeech* (pp. 3414–3418). New York: IEEE.

Michelsanti, D., Tan, Z.-H., Zhang, S.-H., Xu, Y., Yu, M., Yu, D., and Jensen, J. 2021. An overview of deep-learning-based audio-visual speech enhancement and separation. In *IEEE/ACM Transactions on Audio, Speech, and Language Processing*. New York: IEEE.

Morrone, G., Michelsanti, D., Tan, Z.H., and Jensen, J. 2021. Audio-visual speech in painting with deep learning. In *ICASSP 2021–2021 IEEE International Conference on Acoustics, Speech and Signal Processing (ICASSP)* (pp. 6653–6657). New York: IEEE.

Mudaliar, N.K., Hegde, K., Ramesh, A., and Patil, V. 2020. Visual speech recognition: A deep learning approach. In *2020 5th International Conference on Communication and Electronics Systems (ICCES)* (pp. 1218–1221). New York: IEEE.

Noda, K., Yamaguchi, Y., Nakadai, K., Okuno, H.G., and Ogata, T. 2015. Audio-visual speech recognition using deep learning. *Applied Intelligence*, *42*(4), pp. 722–737.

Oneață, D., Caranica, A., Stan, A., and Cucu, H. 2021. An evaluation of word-level confidence estimation for end-to-end automatic speech recognition. In *2021 IEEE Spoken Language Technology Workshop (SLT)*, pp. 258–265. New York: IEEE.

Petridis, S., Li, Z., and Pantic, M. 2017. End-to-end visual speech recognition with LSTMs. In *2017 IEEE International Conference on Acoustics, Speech and Signal Processing (ICASSP)* (pp. 2592–2596). New York: IEEE.

Rahmani, M.H., Almasganj, F., and Seyyedsalehi, S.A. 2018. Audio-visual feature fusion via deep neural networks for automatic speech recognition. *Digital Signal Processing*, *82*, pp. 54–63.

Sadeghi, M., Leglaive, S., Alameda-Pineda, X., Girin, L,. and Horaud, R. 2020. Audio-visual speech enhancement using conditional variational auto-encoders. *IEEE/ACM Transactions on Audio, Speech, and Language Processing*, vol. 28 (pp. 1788–1800). New York: IEEE.

Thanda, A., and Venkatesan, S.M. 2016. Audio visual speech recognition using deep recurrent neural networks. In *IAPR Workshop on Multimodal Pattern Recognition of Social Signals in Human–computer Interaction* (pp. 98–109). Cham: Springer.

Yang, C.-H.H., Qi, J., Yen-Chi Chen, S., Chen, P.-Y., Marco Siniscalchi, S., Ma, X., and Lee, C.-H. 2021. Decentralizing feature extraction with quantum convolutional neural network for automatic speech recognition. In *ICASSP 2021–2021 IEEE International Conference on Acoustics, Speech and Signal Processing (ICASSP)*, pp. 6523–6527. New York: IEEE.

Yu, W., Zeiler, S., and Kolossa, D. 2021. January. Multimodal integration for large-vocabulary audio-visual speech recognition. In *2020 28th European Signal Processing Conference (EUSIPCO)* (pp. 341–345). New York: IEEE.

Zimmermann, M., Ghazi, M.M., Ekenel, H.K., and Thiran, J.P. 2016. Visual speech recognition using PCA networks and LSTMs in a tandem GMM-HMM system. In *Asian Conference on Computer Vision* (pp. 264–276). Cham: Springer.

6 Deep Learning Models for Speech Recognition

LEARNING OUTCOMES

After reading this chapter, you will be able to:

- Understand the basics of conventional speech recognition techniques.
- Demonstrate an understanding of the techniques of statistical acoustic modeling.
- Understand the basic concepts behind cutting-edge deep-learning-based ASR system.

6.1 TRADITIONAL METHODS OF SPEECH RECOGNITION

For more than 60 years, ASR has been a subject of study. The transcription industry has evolved tons over the past 10 years. However, in practical acoustic settings, accurately understanding spoken utterances remains a challenge. Traditionally, speech recognition process is divided on the basis of two categories as shown in Figure 6.1.

- Based on words
  - Single Word Recognizer: Single word recognizers are used to transcribe spoken isolated words. It is very simple, and HMM-based statistical models are widely accepted for single word recognizers.
  - Continuous Word Recognizer: It has a complex structure.

- Based on speaker

  - Speaker Dependent: The speech of specific speaker is recognized. It requires less training data.
  - Speaker Independent: It is generalized for all speakers, and it requires large training data.

Some of the conventional ASR system discussed in this chapter are as follows:

- Hidden Markov models
- Gaussian Markov Model
- Artificial Neural Network
- Deep Belief Neural Network

The following section deals with the brief discussion of conventional speech recognition models.

DOI: 10.1201/9781003348689-6

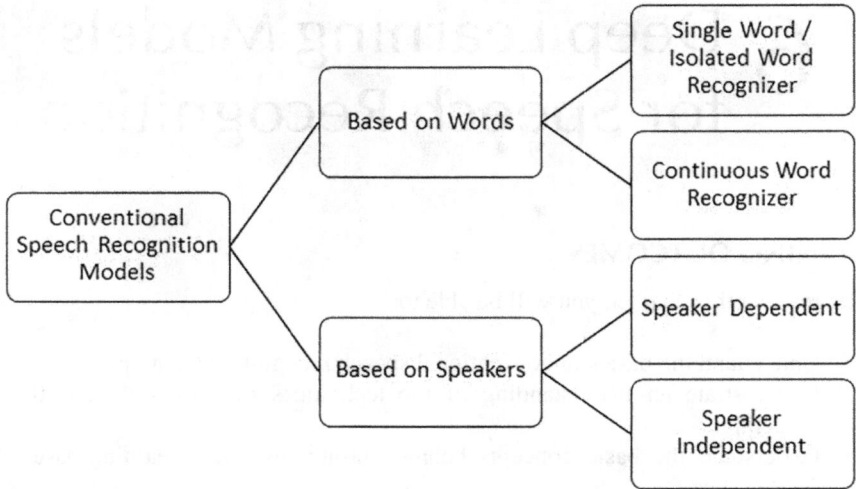

FIGURE 6.1   Conventional Speech Recognition Models.

6.1.1 HIDDEN MARKOV MODELS (HMMS)

Hidden Markov Models are succinct and highly reliable acoustic models. Initially, discrete-density HMM was developed in 1970 by Carnegie Mellon University and IBM. Bell Labs developed continuous-density HMMs. They are widely accepted for many sequence problems. Elements of HMM are as follows:

- Number of state $S = \{s_1, s_2, ... s_n\}$
- Number of distinct observations symbol per state
- State transition probability
- Output probability distributions at every state

The main objective of ASR system is to identify the word sequence $Y = y_1, y_2, .. y_n$, given the acoustic data $X = x_1, x_2, ... x_n$. According to the Bayes rule in Eq. (6.1),

$$P(Y|X) = \frac{P(X|Y).P(y)}{P(X)} \tag{6.1}$$

where

$P(X|Y)$ is the HMM-based acoustic model
$P(Y)$ is the language model
$P(X)$ is a constant for a complete sentence

The architecture of HMM for ASR system is shown the Figure 6.2. HMMs identify the most probable subword units. By concatenating the subword units using generative models, they find the actual spoken text.

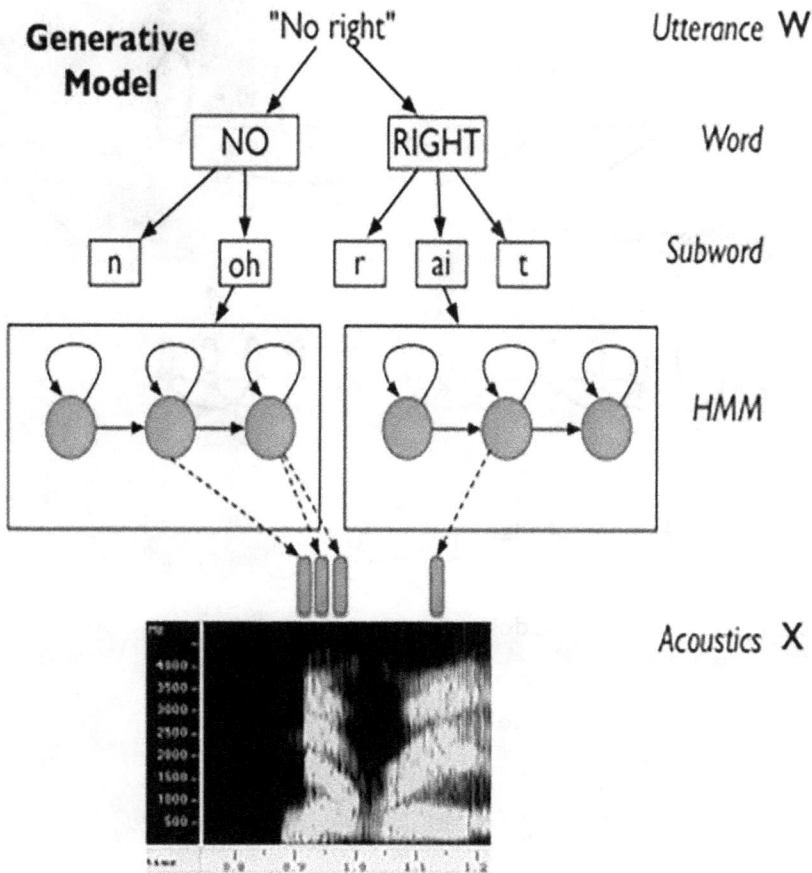

FIGURE 6.2 HMM-based ASR Model.

The types of ASR model which can be trained using HMM are

- Phone-based model: The main idea is to identify the input phonemes (basic sound units) using HMM as shown in Figure 6.3.
- Word-based model: The model, depicted in Figure 6.4, is built by concatenating the sequence of phone models.

The two subtypes of word-based model are:

- Isolated word ASR model: To identify the isolated words, several HMMs are designed and one best model is chosen as shown in Figure 6.5 on the basis of Bayes rule,

$$P(M|X) = P(X|M)p(M)$$

ih

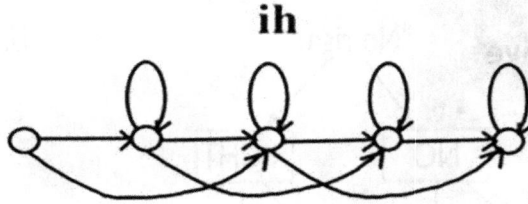

FIGURE 6.3 Phone-based HMM.

d   ih   d

FIGURE 6.4 Word-based HMM.

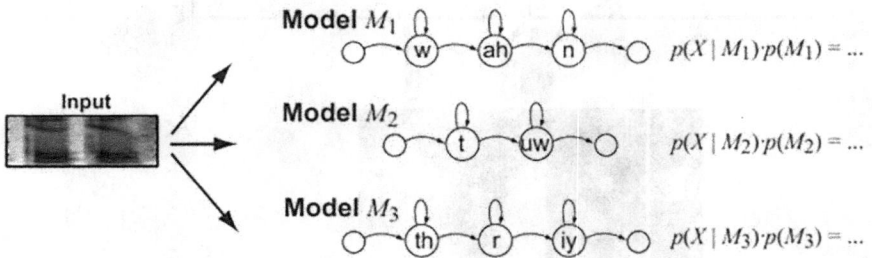

FIGURE 6.5 Isolated Word Recognizer Using HMM.

- Continuous speech ASR model: One large HMM is utilized in identifying spontaneous spoken audio to text as shown in Figure 6.6.

The widely used HMM topologies are shown in figure with their corresponding transition matrix in Figure 6.7.

- Left-to-right model
- Parallel path left-to-right model
- Ergodic model

The three challenges of HMM are as follows:

- Evaluation
  - Problem: To compute the probability of observation sequence, given a model
  - Solution: Viterbi algorithm
- Decoding and alignment

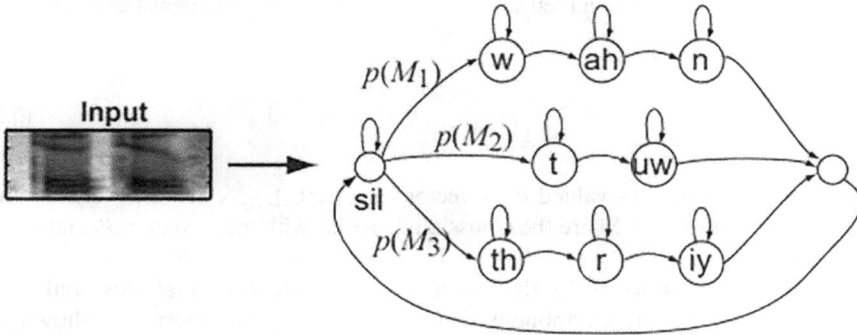

FIGURE 6.6   Continuous Speech Recognizer Using HMM.

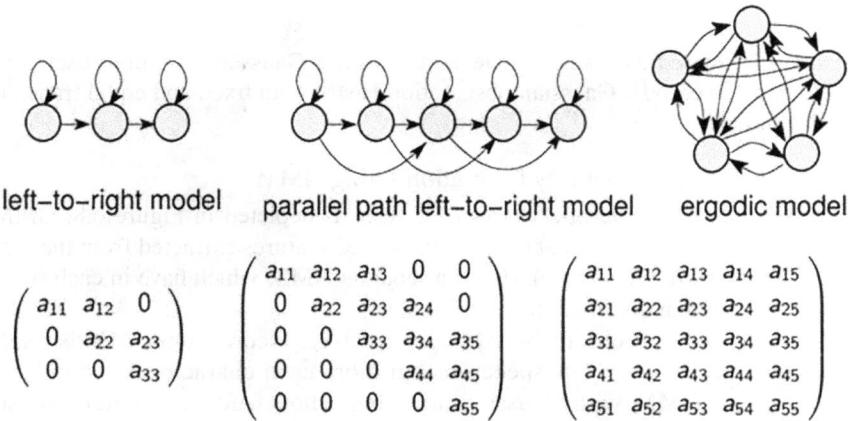

left-to-right model   parallel path left-to-right model   ergodic model

$$
\begin{pmatrix} a_{11} & a_{12} & 0 \\ 0 & a_{22} & a_{23} \\ 0 & 0 & a_{33} \end{pmatrix}
\qquad
\begin{pmatrix} a_{11} & a_{12} & a_{13} & 0 & 0 \\ 0 & a_{22} & a_{23} & a_{24} & 0 \\ 0 & 0 & a_{33} & a_{34} & a_{35} \\ 0 & 0 & 0 & a_{44} & a_{45} \\ 0 & 0 & 0 & 0 & a_{55} \end{pmatrix}
\qquad
\begin{pmatrix} a_{11} & a_{12} & a_{13} & a_{14} & a_{15} \\ a_{21} & a_{22} & a_{23} & a_{24} & a_{25} \\ a_{31} & a_{32} & a_{33} & a_{34} & a_{35} \\ a_{41} & a_{42} & a_{43} & a_{44} & a_{45} \\ a_{51} & a_{52} & a_{53} & a_{54} & a_{55} \end{pmatrix}
$$

FIGURE 6.7   HMM Topologies.

- Problem: To find state sequence which maximizes the probability of observation sequence
- Solution: Viterbi algorithm
- Training
  - Problem: To adjust model parameters to maximize the probability of observed sequences
  - Solution: Forward backward algorithm

6.1.2   GAUSSIAN MIXTURE MODELS (GMMs)

GMM is a statistical based model which utilizes parametric probability density function represented as weighted sum of Gaussian component densities. GMMs are utilized for ASR system to identify the probability densities over the input speech sequence. For training GMMs, they use the estimation algorithm such as Expectation-Maximization (EM) or Maximum A Posteriori (MAP) estimation. A GMM is a weighted sum of M component Gaussian densities as given by Eq. (6.2),

$$p(x/\lambda) = \sum_{i-1}^{M} w_i g(x/\mu_i, \Sigma i)$$ (6.2)

where x is a continuous-valued data vector, $w_i, i = 1, 2, \ldots N$ are the weights, and $g(\zeta| \frac{1}{4}, \Sigma i)$, $i = 1, 2, \ldots M$ are the Gaussian densities with mean vector μ_i and covariance matrix Σi.

GMM is parameterized by the mean vectors, covariance matrices, and mixture weights from all component densities. These parameters are shown in Eq. (6.3)

$$\lambda = \{w_i, \mu_i, \Sigma i\}, i = 1 \ldots\ldots\ldots M$$ (6.3)

GMM is considered as a single-state HMM with a Gaussian mixture observation density or as an ergodic Gaussian observation HMM with fixed and equal transition probabilities.

6.1.2.1 Emission Probability Estimation Using GMM

HMM–GMM is the state-of-art model for ASR as depicted in Figure 6.8. GMMs are the emission probabilities of continuous MFCC features extracted from the input audio. Each word in the vocabulary has a separate HMM which have in each state a mixture of Gaussian distribution.

For a new unseen word, the word with more likely score from HMM is finally selected as the outcome. For speech recognition, each character is individually modeled using HMM with Gaussian mixtures. Though for a complex acoustic model, the phoneme context is taken onto the traditional maximum, the likelihood (ML) estimation is used for estimating the parameter of HMM–GMM acoustic.

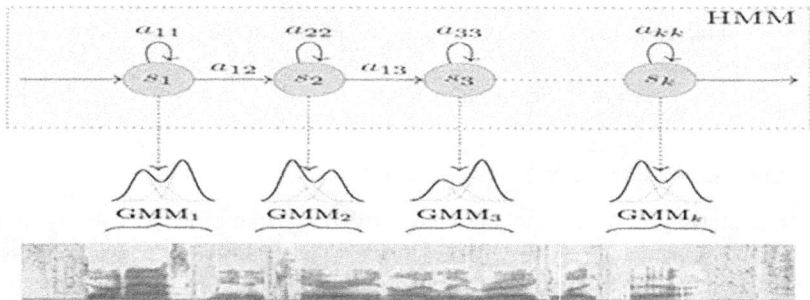

FIGURE 6.8 ASR Using HMM and GMM.

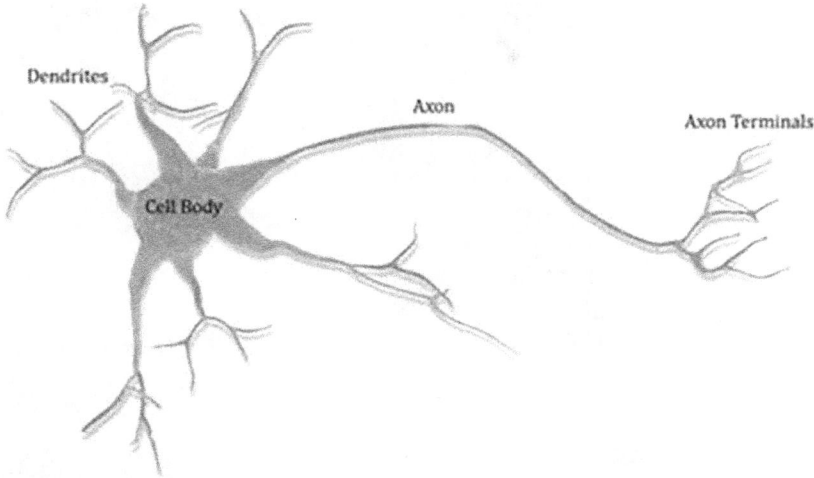

FIGURE 6.9 Biological Neuron.

6.1.3 ARTIFICIAL NEURAL NETWORK (ANN)

ANN is motivated from the working of the human brain. A typical biological neural network is shown in Figure 6.9. The human brain is composed of multiple interlinked neurons and forms complex neural structures to transfer information.

In 1871, Joseph von Gerlach proposed that the nervous system is a single continuous network and is composed of a network of many discrete cells. The term neuron was coined by Hein rich Gottfried von Waldeyer-Hartz around 1891. In 1953, McCulloc (a neuroscientist) and Pitts (logician) proposed a simple model of artificial neuron as shown in Figure 6.10. It takes input from various sources $\{x_1, x_2, \ldots x_n\}$, and the output is a binary decision.

Later around 1956, Frank Rosenbeltt proposed the perceptron model as depicted in Figure 6.11: "the perceptron may eventually be able to learn, make decisions and translate languages."

Perceptron for binary classification, which is used to separate two classes $\{0,1\}$ is represented in Eq. (6.4),

$$y(x_1, x_2, \ldots, x_n) = f(w_1.x_1 + \ldots + w_n.x_n)$$

(6.4)

where x_i is the input and is the learned weight to map input $x_i \rightarrow y_i$.

Multilayered perceptron consists of one input and one output layer with at least one hidden layer as shown in Figure 6.12.

Four steps involved to train a neural network are as follows:

- Forward backward propagation
- Weight updation
- Model optimization

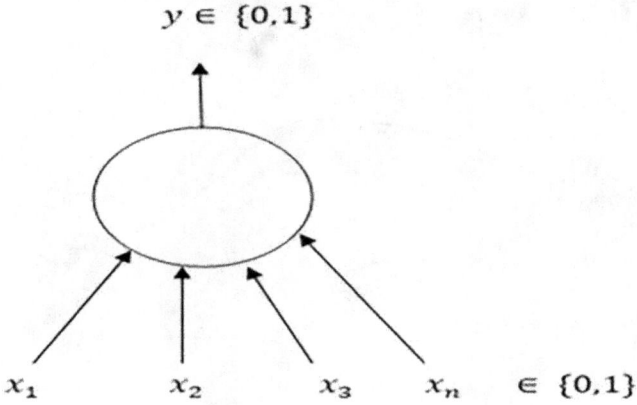

FIGURE 6.10 Simple Artificial Neuron.

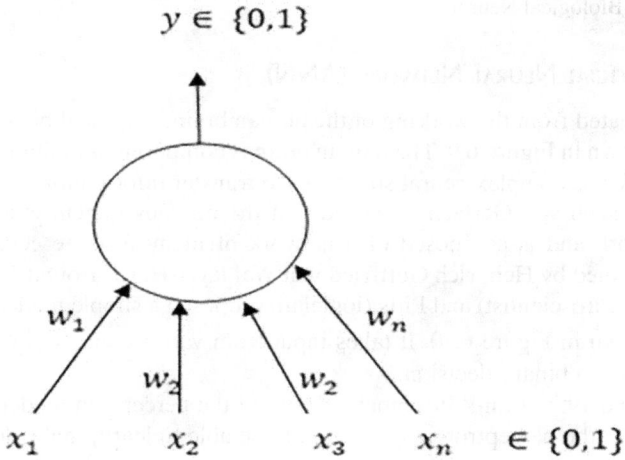

FIGURE 6.11 Perceptron.

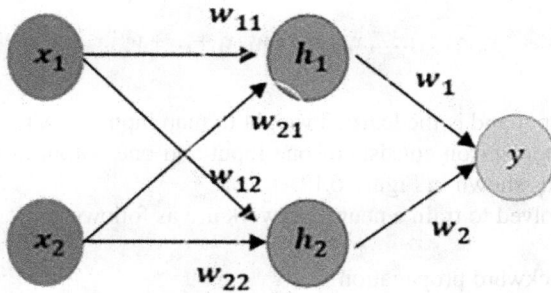

FIGURE 6.12 Multilayered Perceptron.

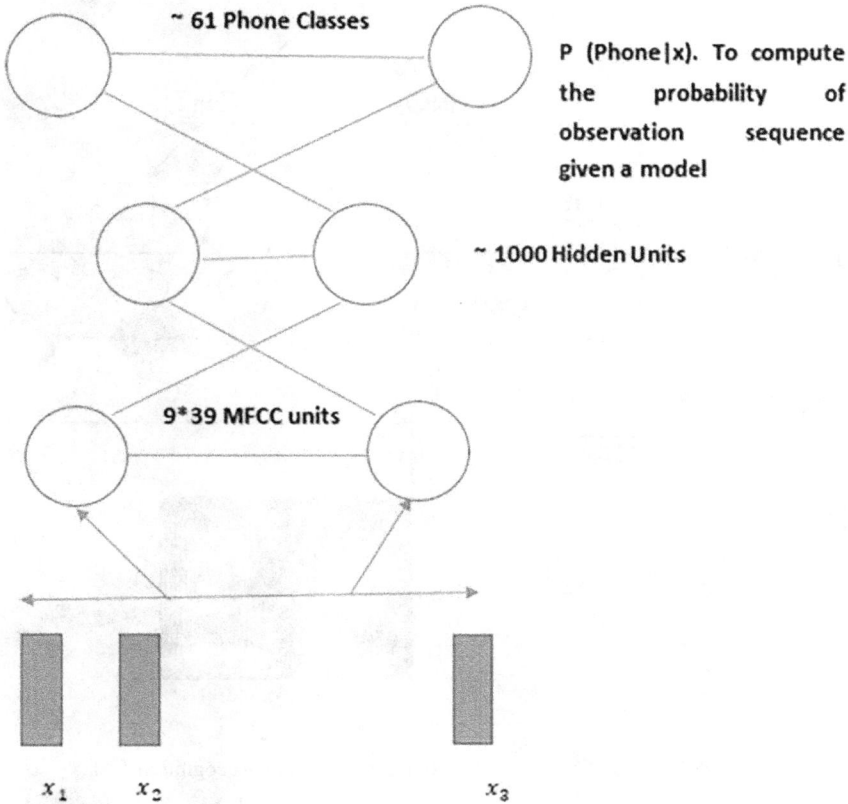

~ 61 Phone Classes

P (Phone|x). To compute the probability of observation sequence given a model

~ 1000 Hidden Units

9*39 MFCC units

x_1 x_2 x_3

FIGURE 6.13 ANN-based Phoneme Classification.

Speech is a continuous signal which can be recognized using neural networks. For handling, the time series and the temporal relationship of acoustic signal, neural networks such as recurrent neural networks (RNN), long short-term memory have been widely used. Neural-network-based ASR system gives a promising result for continuous speech recognition.

A simple ANN for recognition phonemes is depicted in Figure 6.13. The input sequence is $X = \{X_1, X_2, \ldots X_t\}$ whose MFCC features are extracted and fed into the hidden layer of approximately 1,000 hidden units. There are approximately 61 phonemes as the output classes. Given the acoustic feature sequence, ANN predicts the probable phoneme sequence.

$$Phoneme\ Recognition\ Task : P(phoneme \mid acoustic\ feature) \qquad (6.5)$$

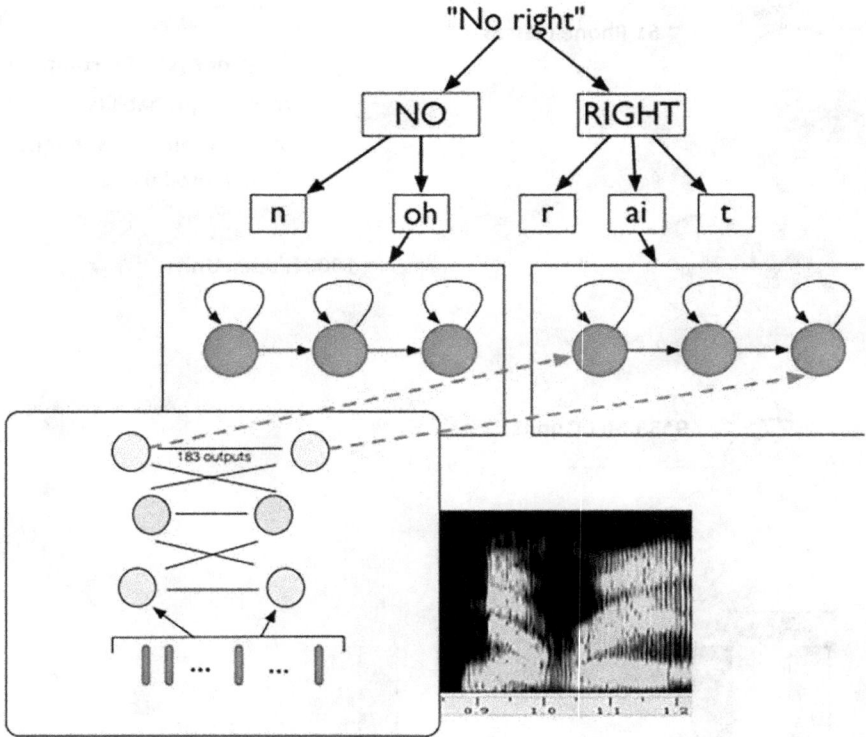

FIGURE 6.14   HMM and ANN Acoustic Modelling of Phone Recognition Task.

6.1.4   HMM AND ANN ACOUSTIC MODELING

The hybrid HMM and ANN model is used to recognize the sequence of phonemes. ANN is used to identify the phonemes from the input acoustic feature. The most likely and probable phoneme sequences are modelled using posterior probabilities of HMM as shown in Figure 6.14. Also, HMM captures the temporal variability of speech in an efficient manner.

In the study, HMM and neural network architecture are combined to identify the spoken utterances as shown in Figure 6.15. Neural network model is designed using ReLU activation, Maxout unit, and Euclidean norm functions. The output from NN is fed into HMM which has scaled likelihood and posterior probability factors. Hybrid HMM with neural network model performs better and in a more flexible way when compared with GMMs.

6.1.5   DEEP BELIEF NEURAL NETWORK (DBNN) FOR ACOUSTIC MODELLING

Acoustic modelling is well known which is a combination of hidden Markov model and artificial neural network. It utilizes both supervised and unsupervised learning

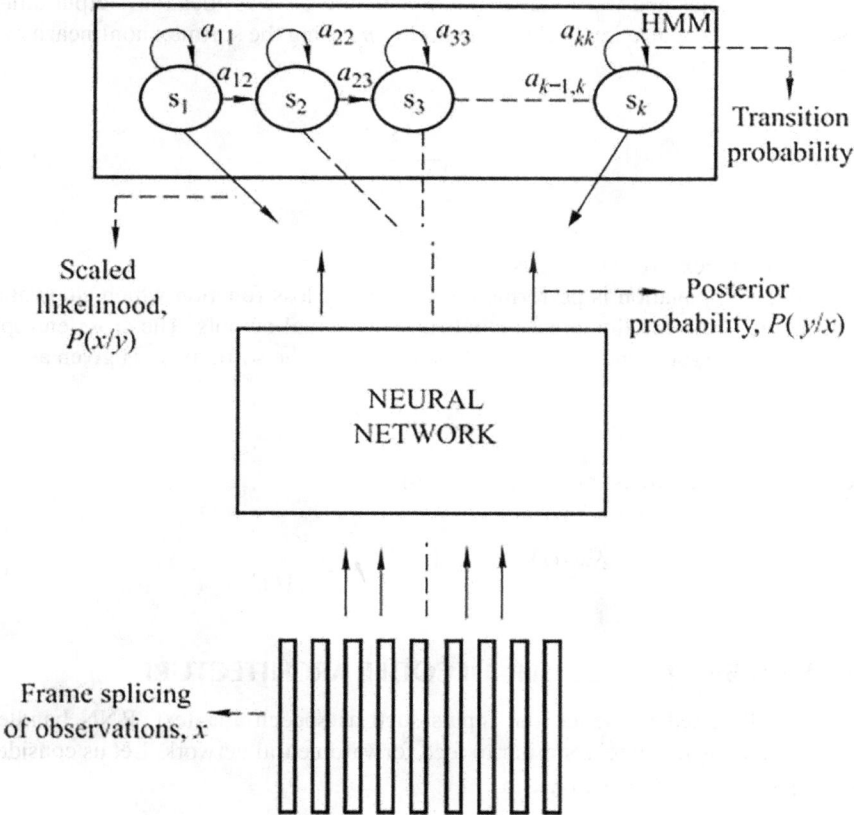

FIGURE 6.15 HMM and ANN Acoustic Modeling of ASR.

for feature extraction, feature selection, and optimization process. DBNN-based ASR modeling efficiently handles the nonlinearity of speech signals. It is composed of a stack of neural layers. DBNN is a graphical model with multiple layers of binary latent variables that can be learned effectively, one layer at a time, by utilizing an unsupervised learning method that maximizes a variation lower bound on the log probability of the acoustic signals. Each hidden unit j uses the logistic function. The logistic function is shown in Eqs. (6.6) and (6.7),

$$y_j = logistic(x_j) = \frac{1}{1+e^{-x_j}} \tag{6.6}$$

and

$$x_j = b_j + \sum_i y_i w_{ij} \tag{6.7}$$

where b_j is the bias and w_{ij} is the weight. For multiclass classification, output unit j converts its total input x_j into a class probability p_j, using the softmax nonlinearity as shown in Eq. (6.8).

$$p_j = \frac{exp(x_j)}{\sum_k exp(x_k)} \qquad (6.8)$$

where k is an index over all classes.

The back propagation is performed to reduce the loss function which calculates the difference between the target outputs and the actual outputs. The cross-entropy c between the target probabilities d and the outputs of the softmax, p, is given as:

$$c = -\sum_j d_j log p_j \qquad (6.9)$$

The weight update rule is shown in Eq. (6.10).

$$\Delta w_{ij}(t) = \alpha \Delta w_{ij}(t-1) - \epsilon \frac{\partial C}{\partial w_{ij}(t)} \qquad (6.10)$$

6.2 RNN-BASED ENCODER–DECODER ARCHITECTURE

RNN is well suited for sequential inputs such as speech and text. RNN handles sequential data which works similar to feed forward neural network. Let us consider the input sequence $X = \{x_1, x_2, \ldots, x_n\}$

$$y_t = f(x_t, h_{t-1}) \qquad (611)$$

where the function f maps the input x_t to output y_t h_{t-1} is the memory from previous input.

- Gated recurrent unit (GRU): It is composed of two gates and the information flows unidirectionally:
  - Reset gate
  - Update gate
- Bi(GRU): The information is carried out bidirectionally through the network of GRU units.
- Long Short Term Memory (LSTM) unit: It is a widely used neural network architecture which is composed of three gates as follows:
  - Input gate
  - Forget gate
  - Output gate
- Bi-LSTM: Bidirectional LSTM unit where the information flows in both the directions.

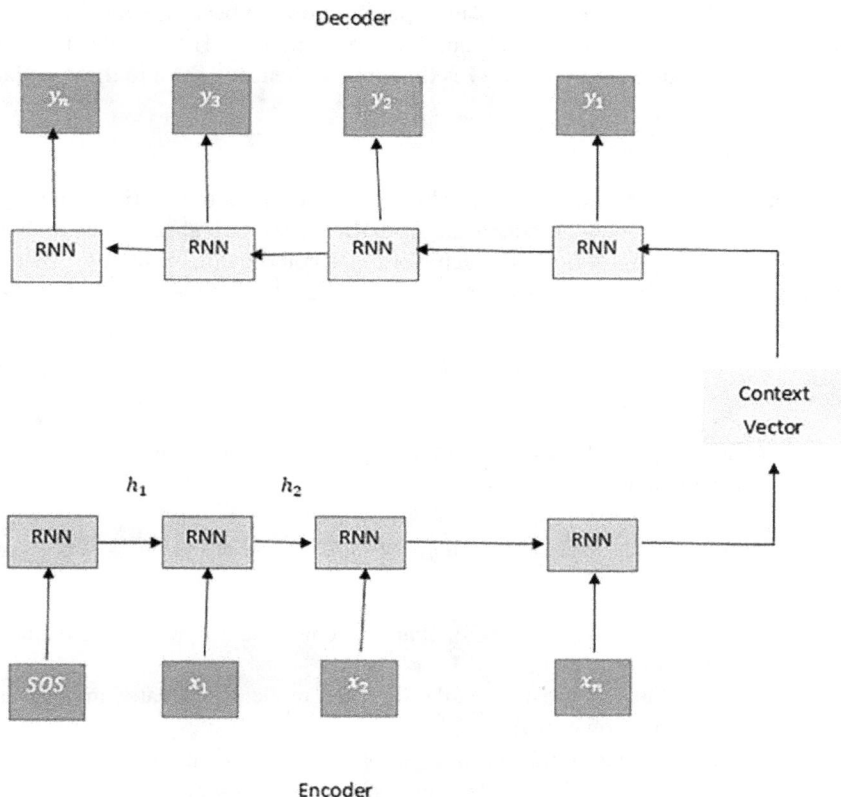

FIGURE 6.16   Encoder–Decoder Structure with RNN Units.

The input sequence is fed into encoder–decoder architecture which is composed of RNN units as depicted in Figure 6.16.

6.3 ENCODER

The input speech sequence consists of continuous speech frames $\{x_1, x_2, \ldots\ldots x_n\}$. The input sequences are then sent to a group of several RNN cells such as LSTM or GRU units for processing the sequential data, each of which admits one input sequence element, gathers data for that element, and transmits that data forward. The hidden states h_i where the processing occurs are calculated using Eq. (6.12).

$$h_t = f\left(W^{(hh)}h_{t-1} + W^{(hx)}x_t\right)$$ (6.12)

where h_t is the hidden layer, W is the weights, and h_{t-1} is the previous hidden layer.

The values of the weights are updated appropriately by back propagation algorithm to the previous hidden state h_{t-1} and the input vector x_t. The context from the encoder part which contextually analyzes the input sequence is fed into the decoder.

6.4 DECODER

A group of several RNN units, each of which estimates an output at time step t, y_t. Each recurrent unit receives a hidden state from the prior unit and generates both an output and a hidden state of its own. Each word is denoted by the symbol y_i, where i denotes the word's order. The hidden state h_i is computed using the Eq. (6.13).

$$h_t = f\left(W^{(hh)}h_{t-1}\right)$$
(6.13)

Using the hidden state at the current time step and the appropriate weight W, the output (y_t) is calculated.

$$y_t = softmax\left(W^S h_t\right)$$
(6.14)

Using softmax, the probability vector that will enable us to predict the result is calculated as shown in Eq. (6.14).

The performance of RNN-based encoder–decoder model is evaluated in WER on LibriSpeech dataset as shown in Table 6.1.

The RNN-based encoder model is suitable for clean speech, and it can be further improved if it is concatenated with language models like n-gram and BERT. But when the length of input speech is longer, RNN model is not suitable. To overcome the challenges attention-based models are used which is discussed in the following section.

6.5 ATTENTION-BASED ENCODER–DECODER ARCHITECTURE

Attention models are introduced by Ian Goodfellow, which is one of the cutting-edge researches happening in the field of deep learning. Attention-based models handle the longer-input speech sequence and effectively analyze the context of the audio

TABLE 6.1
Performance Analysis of RNN-based Encoder–Decoder Model

LibriSpeech Dataset	WER (in %)
Dev-Clean	5.9
Test-Clean	8.5
Test-Other (Noisy)	13.1

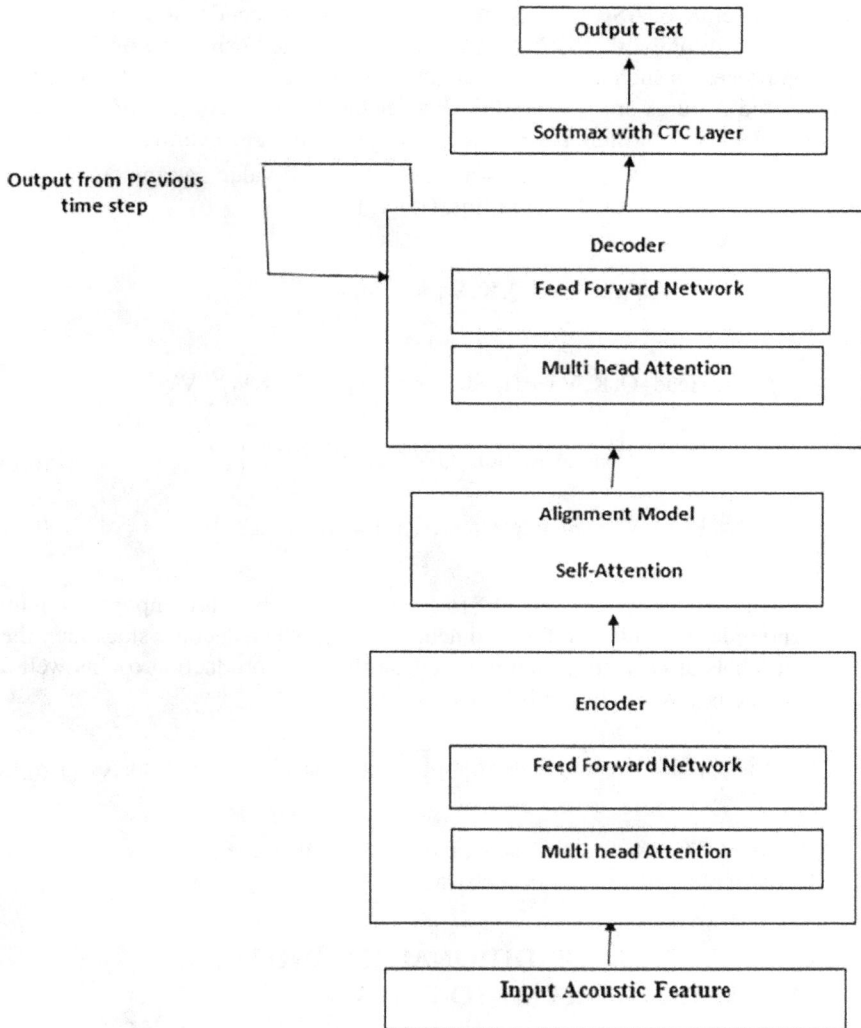

FIGURE 6.17 Encoder–Decoder Structure with Attention Units.

samples. Attention units are used in the encoder–decoder modules of ASR system. The various attention units used are as follows:

- Self-attention: Attention to itself.
- Multi-head attention: Stacks of multiple self-attention units are known as multi-head attention.

More about the working of attention units is discussed in Section 3.4 of Chapter 3.

The architecture of ASR with attention-based encoder decoder structure is shown in Figure 6.17. Initially, the acoustic features are extracted using one of the feature extraction processes such as spectrogram, MFCC, and log mel filter bank energies. The extracted features are fed into the encoder block which consists of multi-head attention layer with normalization followed by feed forward neural networks. The input acoustic vector $(i_1, i_2 ..., i_t)$ acts as query, key, and value parameters and produces the context vector as shown in Eqs. (15–17):

$$\text{Attention}(Q, K, V) = \text{Softmax}\frac{QK^T}{\sqrt{d_k}}V \qquad (6.15)$$

$$\text{Multi-Head}(Q, K, V) = \text{Head}_1\text{Attention}\left(QW_1^Q, KW_1^Q, VW_1^Q\right)\cdots$$

$$\text{Head}_1\text{Attention}\left(QW_n^Q, KW_n^Q, VW_n^Q\right) \qquad (6.16)$$

$$CV_a = \text{encoder}(i_1, i_2 ..., i_t) = [a_1, a_1 ..., a_t] \qquad (6.17)$$

The output of the encoder, CV_a, is fed into decoder block which is composed of multi-head attention layers with feed forward neural network. The decoder stack identifies the output labels at each time step o_t based on the prior production o_{t-1} as well as context vector as given in Eq. (6.18).

$$\text{Output label at current time step } o_t = \prod_t p(o_t|\text{fused context vector}, o_{t-1}) \quad (6.18)$$

Finally, the CTC loss function is applied which makes use of a special character called blank to remove the duplicate characters.

6.6   CHALLENGES IN TRADITIONAL ASR AND THE MOTIVATION FOR END-TO-END ASR

Challenges are listed next based on various factors, and many of them are still not addressed.

- Based on acoustics and phonetics:
  - Ambient noise
  - Age of the speaker
  - Dialects
  - Accent

- Linguistics: Size of vocabulary and word variations
  - To develop acoustic and language models well, lexicon table is a prominent requirement, which depends on a handcrafted pronunciation

dictionary. Such man-made dictionaries are subject to errors. Also, many low-resource languages don't have proper scripting.
- Conditional independence assumptions
  - Real-world speech data do not possess any conditional independence assumptions. Models which use conditional independence assumption may lead to errors.
- Complex decoding:
  - Integrating the modules is harder, which can be resolved by the usage of transducers. Such transducer development and optimization are tough processes.
- Incoherence in optimization:
  - Optimizations of three modules are carried out individually, which can lead to incoherence in optimization.

These are a few of the unresolved research problems that need to be solved. And following this, many researchers and leading tech giants have developed cutting-edge end-to-end pre-trained ASR architectures which are discussed in Chapter 7.

6.7 SUMMARY

The various traditional ASR models, including HMM, GMM, and hybrid HMM–DNN, are explored in this chapter. The idea of an encoder–decoder model based on RNN was also analyzed in order to handle the continuous speech signals. Chapter 7 examines the more sophisticated state-of-the-art deep-learning-based pre-trained end-to-end ASR systems to address the drawbacks of these traditional models.

BIBLIOGRAPHY

J. Bilmes (2008). Gaussian models in automatic speech recognition. In: Havelock, D., Kuwano, S., and Vorländer, M. (eds) *Handbook of Signal Processing in Acoustics*. Springer, New York.

J. Bilmes (2003). Buried Markov models: A graphical modeling approach to automatic speech recognition. *Computer Speech & Language*, 17: 213–231.

A. Dutta, G. Ashishkumar and C.V.R. Rao (2021). Performance analysis of ASR system in hybrid DNN-HMM framework using a PWL Euclidean activation function. *Frontiers of Computer Science*, 15: 154705.

J.P. Haton (1999). Neural networks for automatic speech recognition: A review. In: Chollet, G., Di Benedetto, M.G., Esposito, A., and Marinaro, M. (eds) *Speech Processing, Recognition and Artificial Neural Networks*. Springer, London.

L. Rabiner and B. Juang (1986). An introduction to hidden Markov models. *IEEE ASSP Magazine*, 3(1): 4–16.

7 End-to-End Speech Recognition Models

LEARNING OUTCOMES

After reading this chapter, you will be able to:

- Identify the appropriate end-to-end ASR system for your downstream applications.
- Understand the basic concepts behind end-to-end ASR and online streaming ASR system.
- Identify the right open-source ASR model for inferencing.

7.1 END-TO-END SPEECH RECOGNITION MODELS

Before the rise of deep learning, conventional ASR models were complex systems that included acoustic models, language models, and pronunciation models. The main drawbacks of these was that every modules make assumptions on probability distributions. For instance, n-gram language model and HMMs make strong Markovian independence assumptions between tokens in a sentence.

Now, the advancements in deep learning models can overcome the challenges in conventional ASR system as discussed in Chapter 6. In the conventional ASR system, we have separate acoustic, pronunciation, and language models to transcribe the spoken utterances to text as shown in Figure 7.1. Each of these modules needs to be separated and trained to be optimized whereas end-to-end ASR model can replace it with one single neural network architecture.

7.1.1 DEFINITION OF END-TO-END ASR SYSTEM

A system which directly maps a sequence of input acoustic features into a sequence of graphemes or words

(A. Graves and N. Jaitly 2014)

This chapter explores the well-known end-to-end ASR systems such as CTC, Listen Attend and Spell, deep speech 1, and deep speech 2.

7.1.2 CONNECTIONIST TEMPORAL CLASSIFICATION (CTC)

The first end-to-end models were explored by Alex Graves of Google Deep Mind called Connectionist Temporal Classification (CTC). This technique works on the

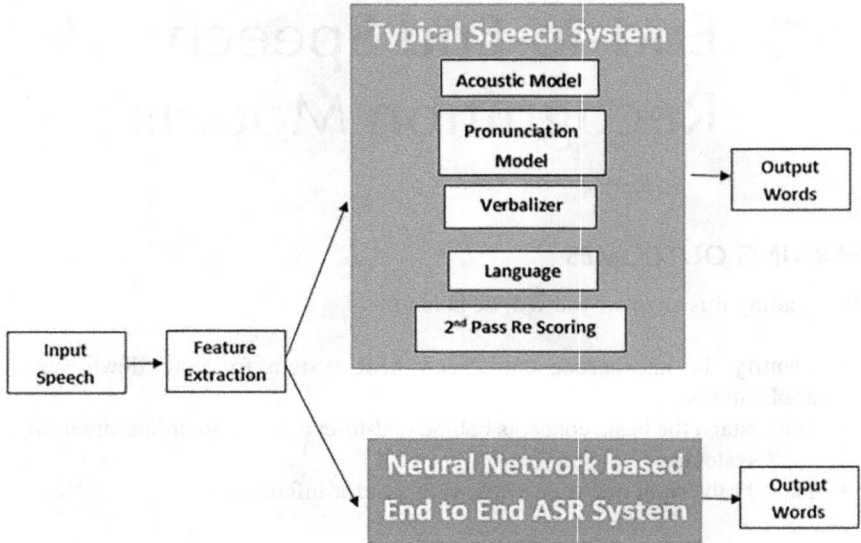

FIGURE 7.1   Conventional versus End-to-End ASR System.

basis of dynamic programming and widely accepted ASR and NLP applications. The main advantage of CTC layer is that no prior alignment between the input and target sequences is required. Alex Graves in developed a standard RNN with CTC for end-to-end ASR systems that, for any given input speech $x = (x_1, x_2, \ldots x_T)$, which is fed into a bidirectional RNN layer as shown in the Figure 7.2, produce the hidden vector $h = (h_1, h_2, \ldots h_T)$ and output vector $y = (y_1, y_2, \ldots y_T)$.

Figure 7.2 shows the bi-RNN which processes both forward and backward hidden vectors as shown in Eqs. (7.1) to (7.3),

$$\overrightarrow{h_t} = H\left(W_{X\vec{h}}x_t + W_{\vec{h}\vec{h}}\overrightarrow{h}_{t-1} + b_{\vec{h}}\right) \tag{7.1}$$

$$\overleftarrow{h_t} = H\left(W_{X\overleftarrow{h}}x_t + W_{\overleftarrow{h}\overleftarrow{h}}\overleftarrow{h}_{t-1} + b_{\overleftarrow{h}}\right) \tag{7.2}$$

$$y_t = W_{\overrightarrow{h}y}\overrightarrow{h}_t + W_{\overleftarrow{h}y}\overleftarrow{h}_t + b_o \tag{7.3}$$

Each output character from the stack of bi-RNN layers is fed into a final output CTC layer. A special character called Blank(B) for null emission is introduced by CTC as shown in Figure 7.3. The blank character is responsible to remove duplicates. Finally, the output from CTC layer is fed into a softmax activation to identify the most likely sequence of transcriptions.

The following are the pros and cons of CTC architecture:

Outputs

Backward Layer

Forward Layer

Inputs

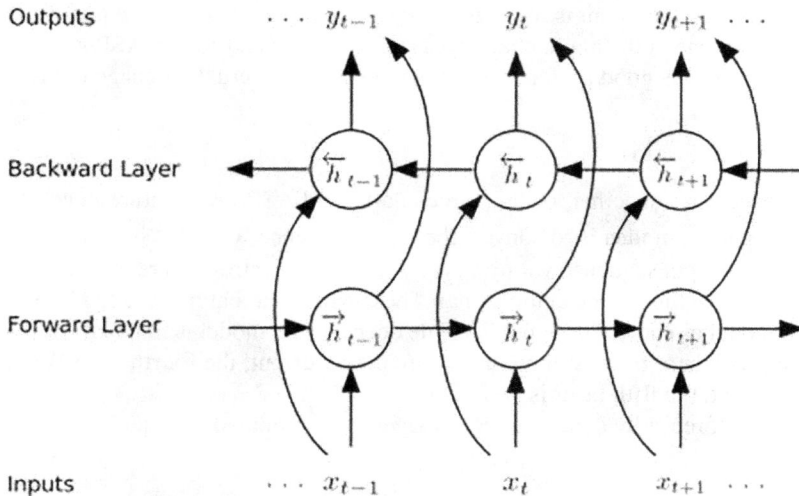

FIGURE 7.2 Bi-RNN.

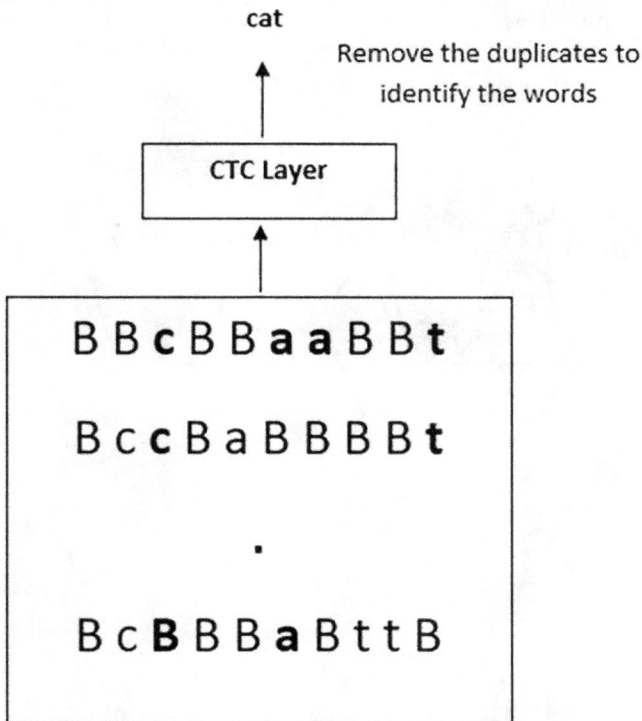

cat

Remove the duplicates to
identify the words

CTC Layer

BBcBBaaBBt

BccBaBBBBt

.

BcBBBaBttB

FIGURE 7.3 CTC.

- **Pros:** No alignment is required between the input and output sequence
- **Cons:** One of the major challenges in CTC-based end-to-end ASR system is that it yields good performance when it uses an external language model

7.1.3 DEEP SPEECH

Deep Speech is a succinct, open-source end-to-end ASR architecture developed by Mozilla. The main idea is to convert the input sequence $x = (x_1, x_2, \ldots x_T)$ to its corresponding output sequence $y = (y_1, y_2, \ldots y_T)$. The spectrogram representations are generated from the source audio signal. These extracted features are fed into a deep speech model as shown in Figure 7.4. The deep speech model is made of five hidden RNN layers where the first three layers are nonrecurrent, the fourth layer is a recurrent layer, and the fifth layer is nonrecurrent.

The first three fully connected RNN layers are computed by Eq. (7.4),

$$h_t^{(l)} = g\left(W^{(l)} h_t^{(l-1)} + b^l\right) \tag{7.4}$$

where $h_t^{(l)}$ is the current hidden layer, $h_t^{(l-1)}$ is the previous hidden layer, W is the weights, and b is the bias.

$$h_t^{(f)} = g\left(W^{(4)} h_t^{(3)} + W_r^{(f)} h_{t-1}^{(f)} + b^{(4)}\right) \tag{7.5}$$

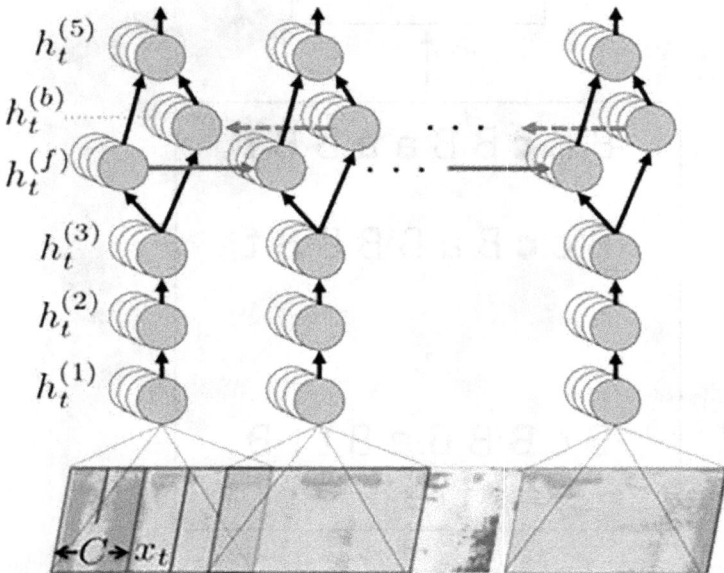

FIGURE 7.4 Deep Speech.

$$h_t^{(b)} = g\left(W^{(4)}h_t^{(3)} + W_r^{(b)}h_{t+1}^{(b)} + b^{(4)}\right) \tag{7.6}$$

where f is forward recurrence and b is backward recurrence. Finally, a CTC layer with softmax activation is utilized to identify the most probable word sequence.

Deep speech model is trained on massive datasets as follows:

- WSJ (*Wall Street Journal*) reads speech with 80 hours.
- Switchboard, continuous speech corpus with 300 hours.
- Fisher, continuous speech corpus with 2,000 hours.
- Baidu, reads speech with 5,000 hours.

Deep speech model achieves a better word error rate when compared with other systems for both clean and noisy speeches as shown in Table 7.1.

7.1.4 Deep Speech 2

Deep speech 2 is the extended version of deep speech 1 in terms of optimization and increased speed. The architecture of deep speech 2 is shown in Figure 7.5.

Steps involved in deep speech 2 architecture are as follows:

- Pre-processing: It converts the raw audio sample into log spectrogram and produces normalized features.
- Model: Deep neural network with two to three convolution layers followed by three to seven GRU/LSTM layers and one fully connected layer. The output labels from the fully connected layer are passed into a CTC layer.
- The main enhancement in deep speech 2 is the utilization of decoder. In deep speech 2, they have considered greedy and beam search decoding mechanism.
- Hyper parameters used are as follows:
  - Learning rate = 0.001
  - Batch size per GPU is 16
  - Stochastic Gradient Descent with momentum is 0.9
  - Batch size per GPU is 16
  - Dropout to lower regularization error

TABLE 7.1

Comparative Analysis of Deep Speech Model

Model	Clean Speech (WER %)	Noisy Speech (WER %)
Apple dictation	14.24	43.76
Google API	6.64	30.47
Deep speech	6.56	19.06

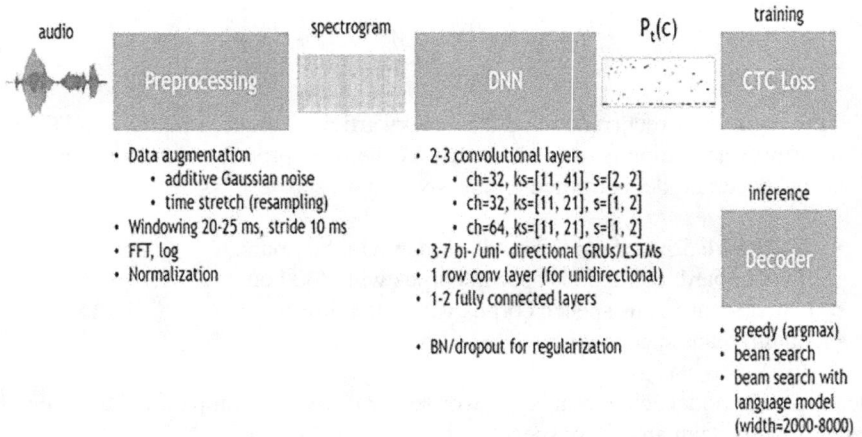

FIGURE 7.5   Deep Speech 2.

TABLE 7.2
Comparison of Deep Speech 2 Model in the Perspective of WER

Dataset	WER (in %)
WSJ	4.42
LibriSpeech (Test Clean)	5.15
LibriSpeech (Test Other)	12.73
Voxforge American Canadian	7.94
Voxforge European	18.44

It is trained on massive amount of data by including LibriSpeech and Voxforge. Overall, it achieves a lowered WER and faster convergence when compared with deep speech 1 as shown in Table 7.2.

7.1.5   LISTEN, ATTEND, SPELL (LAS) MODEL

LAS model is a neural-network-based system with three modules—listen, attend, and spell. An encoder-decoder-based RNN with attention mechanism is utilized in LAS architecture as shown in Figure 7.6. The input is the set of t acoustic features $x = \{x_1, x_2, \ldots, x_t\}$ which is framed with frame size of 10 ms containing 40 dimensional log mel filter bank. Listener is the acoustic model encoder part, attention part is utilized for the alignement between the input and output sequence, and speller is the decoder part as shown in Figure 7.7.

LAS architecture has an encoder (listener), decoder (speller), and attention in between them:

- Encoder (analogous to AM)
  - Transforms input speech x into higher-level representation h. Listener is composed of three pyramidal bidirectional LSTM (pBLSTM). Three

$$P(\mathbf{y}_u | y_{u-1}, \cdots, y_0, \mathbf{x})$$

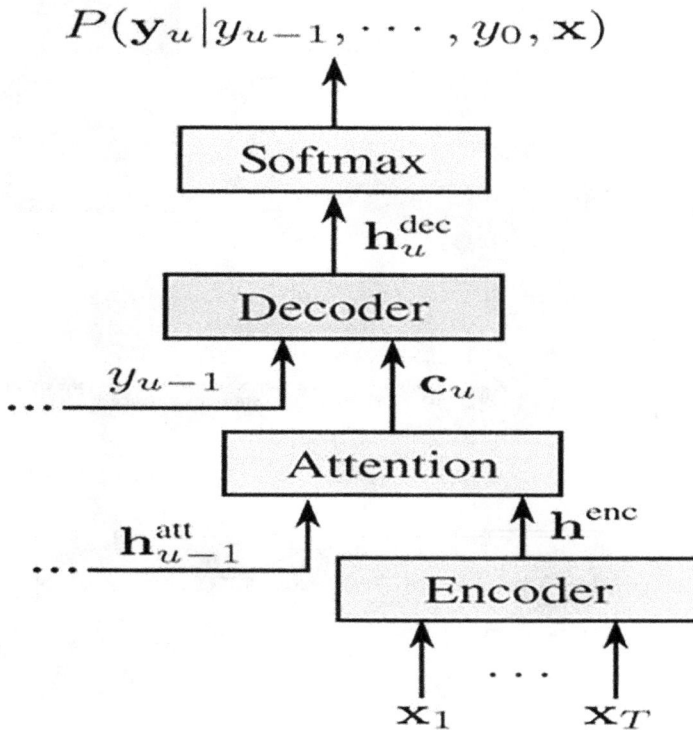

FIGURE 7.6 LAS Architecture.

layers of 512 pBLSTM are stacked on top of LSTM layers to extract the relevant information.

$$h = Encoder\ or\ Listener\ (x) \tag{5.7}$$

$$h_i^j = BLSTM\left(h_{i-1}^j, h_i^{j-1}\right) \tag{5.8}$$

- Attention (alignment model)
  - The output vector is passed into attention-based alignment model. Identifies and performs alignment on encoded frames that are relevant to producing current output.

$$e_{u,t} = score\left(h_{u-1}^{att}, h_t^{enc}\right) \tag{5.9}$$

$$\alpha_{u,t} = \frac{\exp\left(e_{u,t}\right)}{\sum_{t'=1}^{T}\exp\left(e_{u,t'}\right)} \tag{5.10}$$

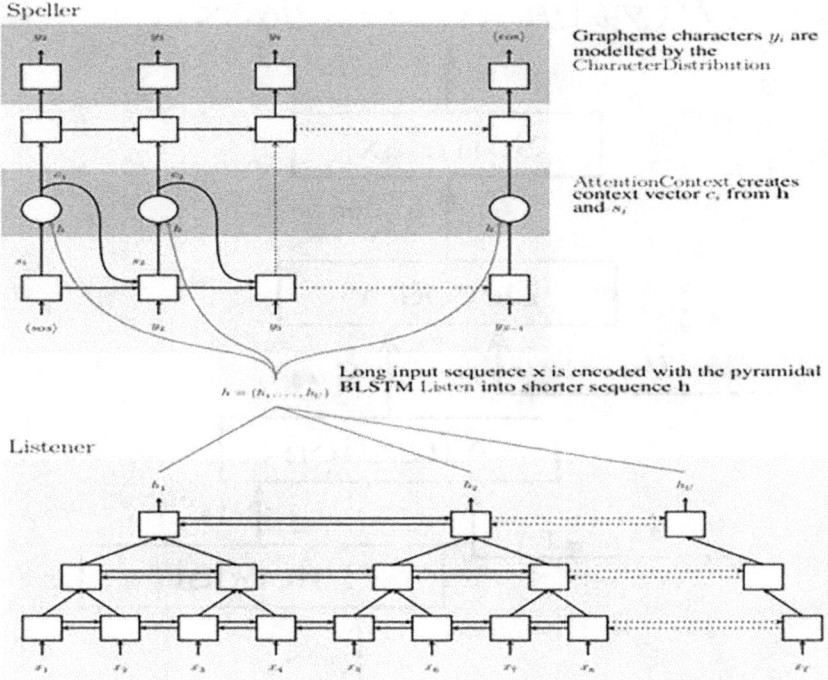

FIGURE 7.7   Listener and Speller.

$$C_u = \sum_{t=1}^{T} \pm_{u,t} h_t^{enc} \tag{5.11}$$

- Decoder (analogous to PM, LM)
  - The last model is the speller part which has two layers of LSTM with 512 nodes each. The current hidden vector along with previous output and previous hidden state is fed into speller to predict the transcription of the input audio.

$$\hat{y}_i \sim Character\ Distribution(s_i, c_i) \tag{5.12}$$

where c_i is the context state and s_i is the decoder state.

LAS model is trained and tested on Google Voice Search utterances which consist of 3M voice searches. It is almost 2,000 hours of data, and they have also augmented the data to increase the dataset size. The LAS model achieved a WER of 14.1% on the clean test set and 16.5% in the noisy set without language model.

7.1.6 JASPER

Jasper (just another speech recognizer) is developed by Nvidia. It makes use of Mel filter bank as input speech feature representation. The window size of each frame is 20 ms with 10 ms shift. Jasper B*R architecture represents B blocks and R subblocks as shown in Figure 7.8. The architecture of sub block is composed of 1D convolution, batch normalization with Rectified Linear Unit as activation and dropout as regularization parameter. Jasper B*R architecture is designed to fast GPU inference. In addition, it possesses four convolution layers, one at the initial stage for pre-processing and other three at the end for post processing. The following are the variants of Jasper:

- Jasper 10*5 is composed of 10 blocks and 5 subblocks. Each subblock consists of five 1D convolutional blocks and four extra blocks.
- Jasper 10*3 architecture is composed of 10 blocks and 3 subblocks. Every sub block is composed of one 1D convolution, batch norm, ReLU, and dropout operations. The following are the details of Jasper 10*3 architecture.

  1. Three variants of normalization batch norm, weight norm, and layer norm are used.
  2. Three variants of rectified linear units ReLU, clipped ReLU (cReLU), and leaky ReLU (lReLU) are used.

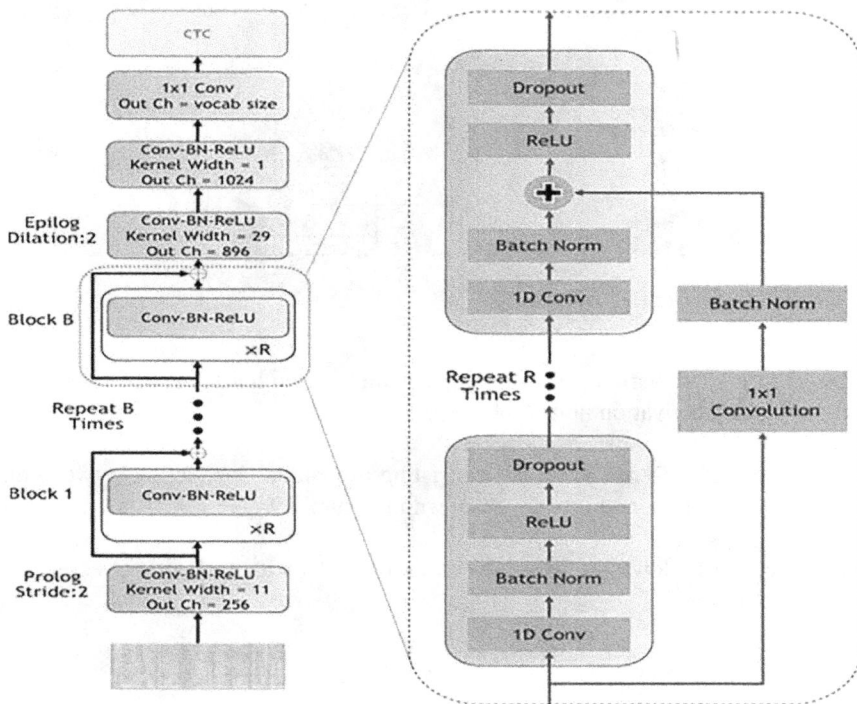

FIGURE 7.8   Jasper B*R Architecture.

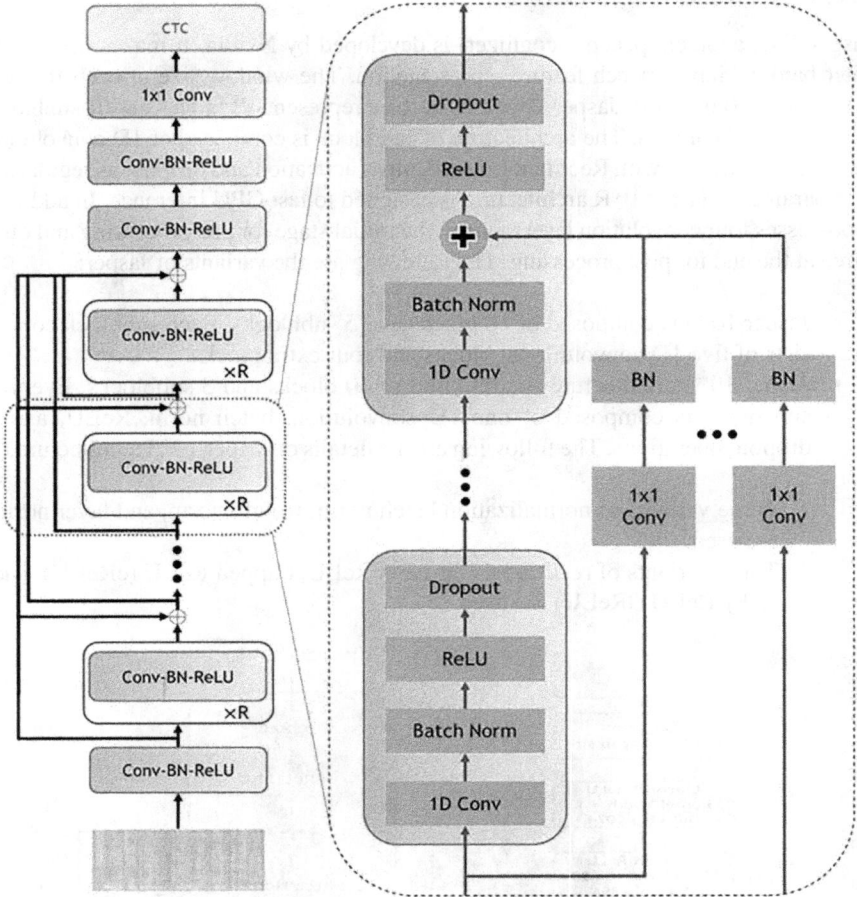

FIGURE 7.9   Jasper Dense Residual.

    3. Uses two variants of gated units, namely gated linear units (GLUs) and
        gated activation units (GAUs).

- Jasper DR (Dense Residual) consisting of residual connection between the
  blocks and the architecture is shown in Figure 7.9.

Jasper achieves a lower WER as shown in Table 7.3 with and without external language model.

7.1.7   QuartzNet

QuartzNet design is variant of Jasper architecture, with convolutional model trained
using Connectionist Temporal Classification (CTC) loss. QuartzNet's architecture

TABLE 7.3

Comparison of Jasper B*R Model

Model	Dataset		Language Model	WER (in %)
Jasper 10×3	LibriSpeech	Dev-Clean	–	4.51
		Dev-Other		4.15
Jasper 10×3	Wall Street Journal	Validation	4-Gram	9.9
		Testing		7.1
		Validation	Transformer-XL	9.3
		Testing		6.9
Jasper 10×5	Hub5'00	Switchboard (SWB)	4-Gram	8.3
		Call Home (CHM)		19.3
		Switchboard (SWB)		7.8
		Call Home (CHM)	Transformer-XL	16.2

is a variant of Jasper got by replacing 1D convolutions with 1D time-channel separable convolutions, an implementation of depth wise separable convolutions. QuartzNet models have the following structure: they start with a 1D convolutional layer followed by a sequence of blocks as shown in Figures 7.10 and 7.11. Each block B_i is repeated S_i times and has residual connections between blocks. Each block B_i consists of the same base modules repeated R_i times and contains four layers that are as follows:

- K-sized depth wise convolutional layer with c_{out} channels.
- A pointwise convolution.
- A normalization layer.
- ReLU.

The following are the parameters:

- Optimizer: NovoGrad
- Learning rate: 0.05
- Two language models: 4 gram and transformer XL

The three variants of QuartzNet are as follows:

- QuartzNet 5*5: It has 6.7 M parameters and it is trained on LibriSpeech corpus for 300 epochs. It achieves a WER of 5.39% with greedy decoder.
- QuartzNet 10*5: QuartzNet 10*5 is composed of 10 blocks and each is repeated for five times.
- QuartzNet 15*5: QuartzNet 15*5 is composed of 15 blocks and each is repeated for five times. It is modelled on datasets like LibriSpeech and Mozilla EN common voice. It achieved a WER of about 4.19% in Libri-Speech with greedy decoder.

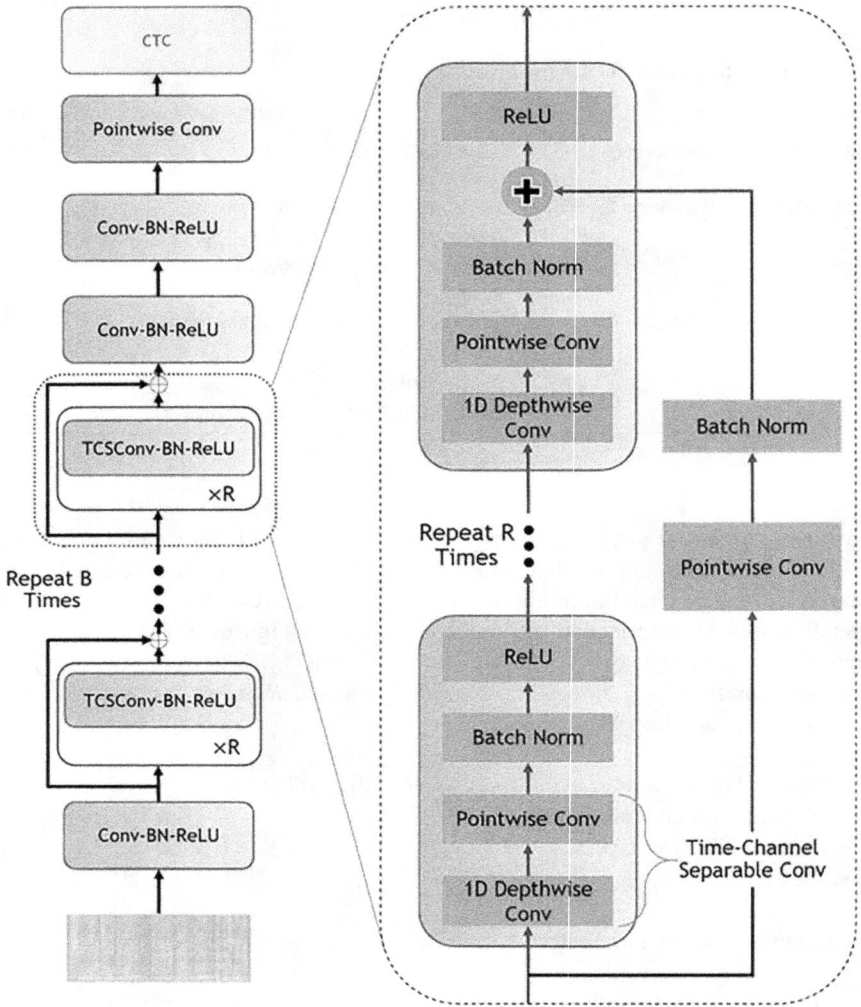

FIGURE 7.10   QuartzNet.

The performance analysis of QuartzNet is compared with that of Jasper and LAS in Table 7.4.

7.2   SELF-SUPERVISED MODELS FOR AUTOMATIC SPEECH RECOGNITION

They are recent state-of-the-art deep learning techniques used to learn unlabeled data. Self-supervised algorithms are extensively accepted for various applications such as speech, text, and image processing. This section explores the more recent ASR architectures such as wav2Vec, data2vec, and hidden unit BERT (HuBERT).

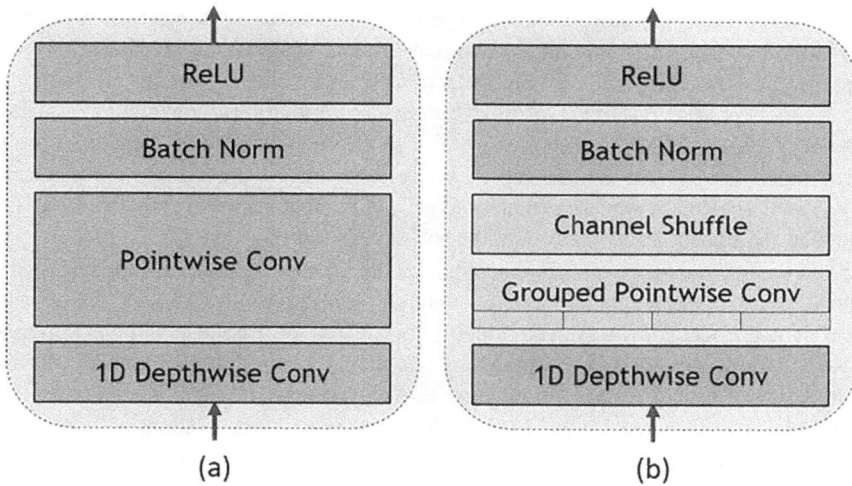

FIGURE 7.11   Quartz Architecture.

TABLE 7.4
Comparison of QuartzNet Model

Model	Language Model	Test Clean (%)	Test Other (%)	Parameters (in Million)
Listen Attend and Spell (LAS)	RNN	2.5	5.8	360
JasperDR 10×5	6-Gram	3.24	8.76	333
	T-XL	2.84	7.84	
QuartzNet 15×5	6-Gram	2.96	8.07	19
	T-XL	2.96	7.25	

7.2.1 WAV2VEC

Wav2Vec 2.0 utilizes a more recent technique called self-supervised learning developed by Meta AI. Self-supervised learning is one of the new deep-learning-based techniques to handle unlabeled data. This model achieves a greater performance in case of varied dialects and more languages. The Wav2vec architecture is composed of following layers:

1. Feature encoder: It takes speech features as input and produces T latent representations. It is composed of multilayered convolution neural network.
2. Transformer: For learning contextually the input sequence and produces a contextualized representation.
3. Quantizer: To produce the target output representation.

Traditional speech recognition models are generally trained using transcriptions of annotated speech audio. Large amounts of annotated data, which is only available for a few languages, are required for good systems. Self-supervision allows you to use unannotated data to improve your systems. Our model consists of a multilayer convolutional feature encoder $f : X \rightarrow Z$ that receives raw audio X as input and produces latent speech representations $Z_1 \ldots Z_T$ for T time steps. They're subsequently passed into a transformer $g : Z \rightarrow C$, which creates representations $C_1, \ldots, C_T$ that capture data from the full sequence. To signify the objectives in the self-supervised objective, the feature encoder output is discretized to q_t with a quantization module $Z \rightarrow Q$. The encoder is made up of multiple blocks that include temporal convolution, layer normalization, and GELU activation. The encoder's raw waveform input is normalized to zero mean and unit variance. The feature encoder's output is routed into a context network that uses the Transformer Architecture. The model uses a multilayer convolutional neural network to analyze the raw waveform of the speech audio to generate latent audio representations of 25 ms each. The quantizer selects a speech unit from an inventory of learned units for the latent audio representation. Before being supplied into the transformer, around half of the audio representations are disguised. The transformer incorporates data from the whole audio track. Finally, the transformer's output is employed to solve a contrastive problem. The model must identify the correct quantized speech units for the masked places in order to complete this task. The architecture of Wav2vec is depicted in Figure 7.12.

The predefined assumption made is to apply mask for particular fraction of the input before the context network as shown in Figure 7.13.

7.2.2 Data2vec

Data2vec is a self-supervised learning approach suitable for multimodal architecture, especially text, speech, and image modalities developed by Meta AI as shown in Figure 7.14.

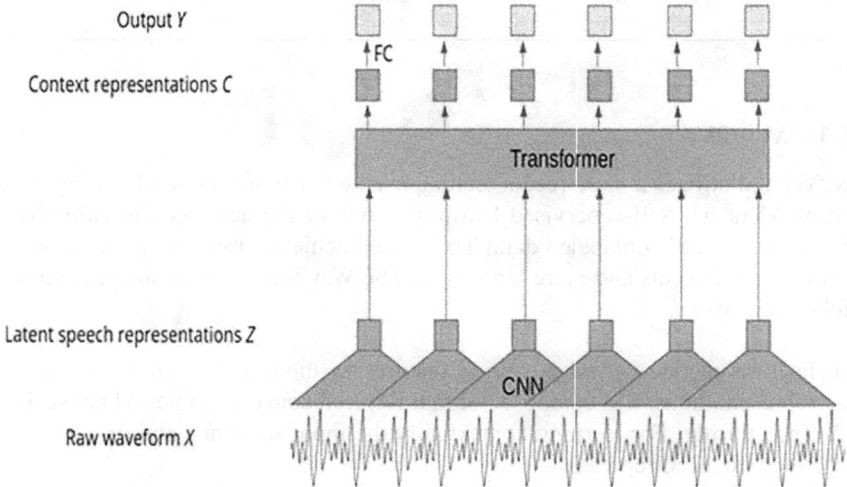

FIGURE 7.12   Wav2Vec Architecture.

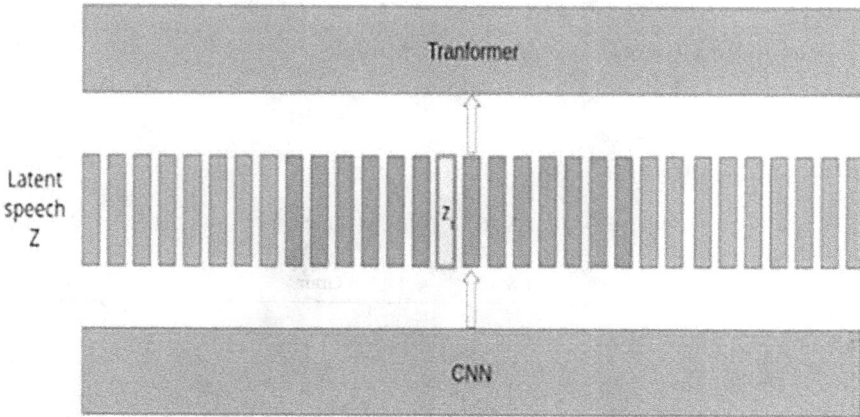

FIGURE 7.13 Wav2Vec with Masked Transformer.

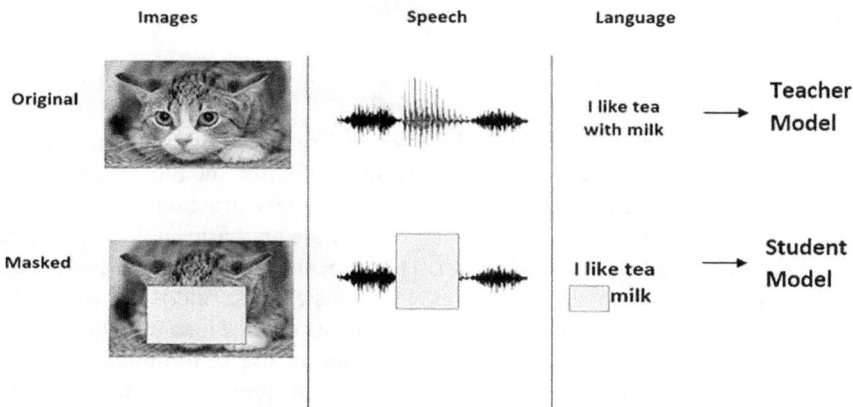

FIGURE 7.14 Data2vec.

Initially, speech signal is encoded with a multilayer 1D convolutional neural network which maps 16 kHz waveform to 50 Hz representations. After the encoding process, masking is applied to the latent speech representation. The masked representation is fed into a transformer-based architecture. The following parameters are used for a data2vec base model.

- Seven temporal convolutions with 512 channels.
- Adam optimizer.
- Learning rate: 5×10^4

Data2vec is trained on LibriSpeech corpus on 960 hours of speech. It is observed from Table 7.5 that it achieves a relative improvement of 20% in WER.

TABLE 7.5

Comparison of Wav2Vec with Data2vec Model

Model	Dataset	Unlabeled Data	Language Model	Labeled Data	
				100 h	960 h
Wav2vec 2.0 (base models)	LibriSpeech	LS-960	4 Gram	8.0	6.1
Wav2Vec 2.0 (large models)	LibriSpeech	LS-960	4 Gram	4.6	3.6
Data2vec	LibriSpeech	LS-960	4 Gram	4.6	3.7

7.2.3   HuBERT

Hidden unit BERT (HuBERT) is a self-supervised ASR model by Meta AI. HuBERT architecture is composed of the following layers as shown in Figure 7.15.

- CNN encoder
- Transformer
- Projection layer
- Code-embedding layer

There are two phases in the modeling, which are generating hidden units and masked prediction. Generating hidden units, that is, finding the hidden units, is the initial phase in the training process, which starts with the extraction of MFCCs (Mel frequency cepstrum) from the audio waveforms. These are basic auditory characteristics that can be used to describe speech. The K-means clustering algorithm is then used to assign each audio segment to one of K clusters. The hidden units are then used to label every audio frame according to the cluster to which it belongs. These units are then transformed into embedding vectors for use in training step B. The output of an intermediary of the BERT encoder from the previous iteration is used by the model to produce representations that are superior to the MFCCs after the first training step.

The second stage, masked prediction, uses masked language modeling to simulate the training of the initial BERT model. Characteristics from the raw audio are created by the CNN and supplied into the BERT encoder after being randomly masked. The masked tokens are filled in by the BERT encoder, which outputs a feature sequence. The cosine similarity between these outputs and each hidden and output embedding created in step A is calculated when this output is projected into a low dimensional space to match the labels. Next, the logits are subjected to the cross-entropy loss to penalize incorrect predictions.

HuBERT variants are as follows.

- HuBERT base
  - No. of CNN encoders is 512.
  - Transformers: 12 layers; dropout probability: 0.05; no of attention heads: 8.

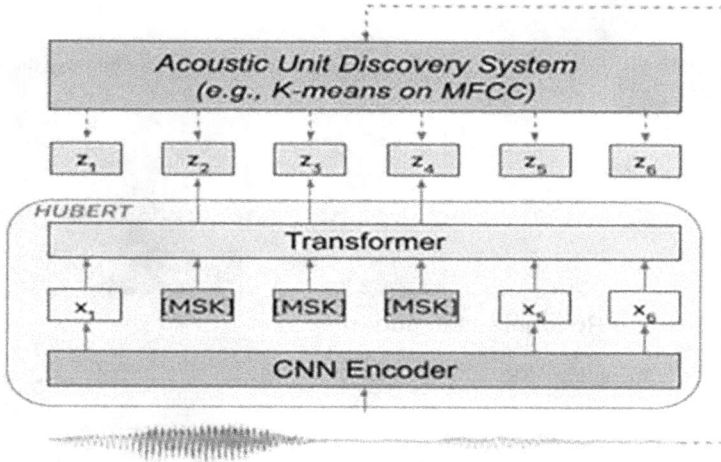

FIGURE 7.15 HuBERT.

- The no. of parameters is 95 M.
- Projection dimension is 256.
- HuBERT large
  - The no. of CNN encoders is 512.
  - Transformers: 24 layers, no of attention heads is 16.
  - No. of parameters is 317 M.
  - Projection dimension is 768.
- HuBERT Xlarge
  - The no/ of CNN encoders is 512.
  - Transformers: 48 layers, no of attention heads is 16.
  - The no. of parameters is 964 M.
  - Projection dimension is 1024.

7.3 ONLINE/STREAMING ASR

Streaming/online ASR is one of the critical application. Streaming ASR is a widely used one in smart speakers, edge device application, live video captioning. One of the most important metrics in online ASR system is to achieve minimum latency. Figure 7.16 shows the working of online ASR-based edge devices. The edge devices which use speech recognition start to recognize the audio once the wake-up word is recognized. The audio or speech that comes after the wake-up word will be taken into account for further processing by the device. Recognition of the audio will be terminated if the device doesn't receive any audio or speech input for a particular time. Then it finalizes the recognition and takes action or fetches the search results from the Internet.

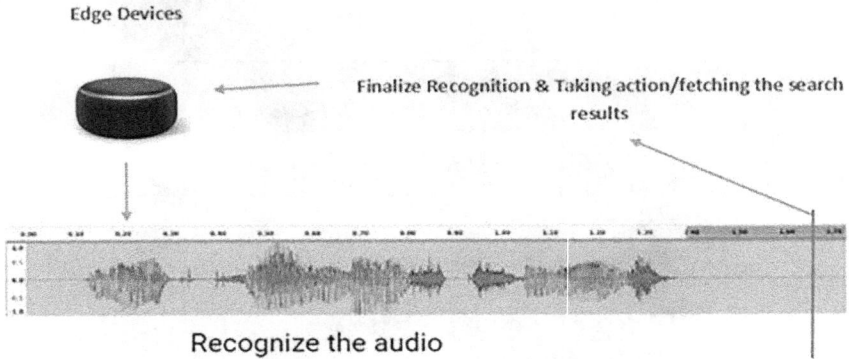

FIGURE 7.16 Streaming ASR.

7.3.1 RNN-TRANSDUCER-BASED STREAMING ASR

RNN-transducer-based streaming ASR system is a variant of sequence-to-sequence models. Unlike offline ASR system, which typically converts the complete input speech to text, the RNN-T continuously processes speech samples and streams output symbols or characters. The RNN-T recognizer outputs characters one-by-one, as you speak, with white spaces where appropriate. Figure 7.17 shows the architecture of RNN-T-based streaming ASR system. The following are some terms describing it.

Encoder network: Set of recurrent layers

Prediction network: Recurrent LM

Joint network: Combines AM and LM predictions—No alignment needed

Table 7.6 shows the RNN-T-based online models' performance in the perspective of WER and EOU (end of utterance) latency.

7.3.2 WAV2LETTER FOR STREAMING ASR

The Wav2Letter model was developed by Meta AI. It consists of 17 1D-convolutional layers and 2 fully connected layers as shown in Figure 7.18.

The acoustic is pre-processed by sampling with a sliding window of 20 ms and a stride of 10 ms. The model's input characteristics are then extracted from these frames as 64-bit log-mel filter bank energies. Connectionist Temporal Classification (CTC) loss is used to train the model. A sequence of letters corresponding to the input speech is the produced output of the model. Clipped ReLU is used to decrease the number of model parameters by roughly half without having an effect on the Word Error Rate (WER). Table 7.7 shows the comparison of wav2letter performance in the perspective of WER.

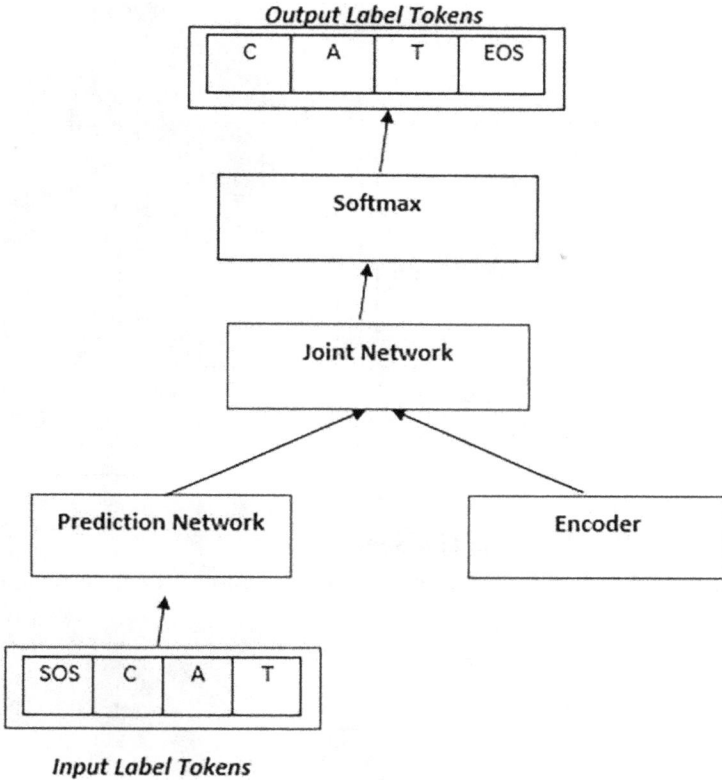

FIGURE 7.17   RNN-T Architecture.

TABLE 7.6
On-Device RNN-T-based Model

Model	Dataset	WER	EOU Latency
On-Device RNN-T +VAD	Voice search	7.4%	860 ms
On-Device RNN-T EP		6.8%	790 ms

7.3.3 CONFORMER MODEL

The CNN and transformer have given a great result in ASR. CNNs effectively use local features, while transformer models identify the global context more efficiently. In conformer models, we have integrated both convolution and transformer models to work in an efficient manner as shown in Figure 7.19.

The conformer consists of two feed forward layers that resemble macrons, sandwiching multi-headed self-attention and convolution modules with half-step residual connections. There is a post layer norm after that.

The conformer encoder consists of multiple blocks to process the audio. Four modules make up a conformer block, which are explained here.

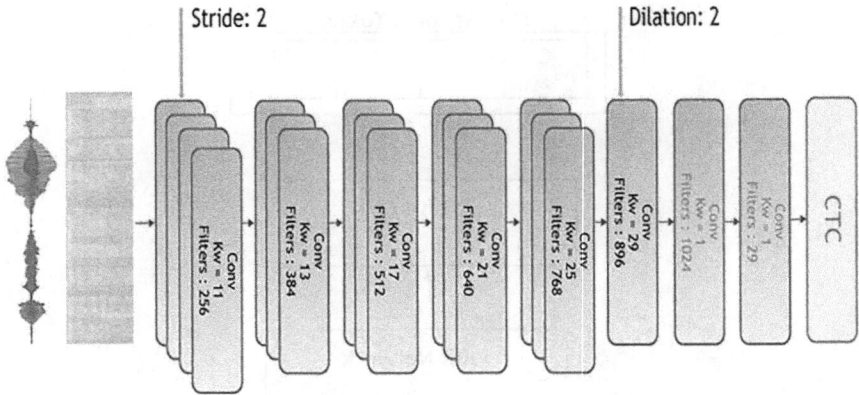

FIGURE 7.18 Wav2Letter Architecture.

TABLE 7.7
Comparison of Wav2Letter in LibriSpeech Corpus

Model	Dataset	WER %, Greedy Decoding	WER %, Beam Search
Wav2Letter	dev-clean	6.67	4.75
	test-clean	6.58	4.94
	dev-other	18.67	13.87
	dev-other	19.61	15.06

- **Multi-headed self-attention module**

Multi-headed self-attention module is used with relative positional embedding, which allows the self-attention model better generalization on different input audios.

- **Convolution module**

Gating mechanism is the starting point of convolution module. One-dimensional depth-wise convolution layer is followed by convolution module.

- **Feed forward module**

Feed forward model is added with a residual connection followed by normalization. The feed forward model has "Swish activation" followed by dropout layer and linear layer.

- **Conformer block**

The multi-headed self-attention module and the convolution module are present between two feed forward modules of the conformer block. Two half feed forward models are present, one before the attention and one after.

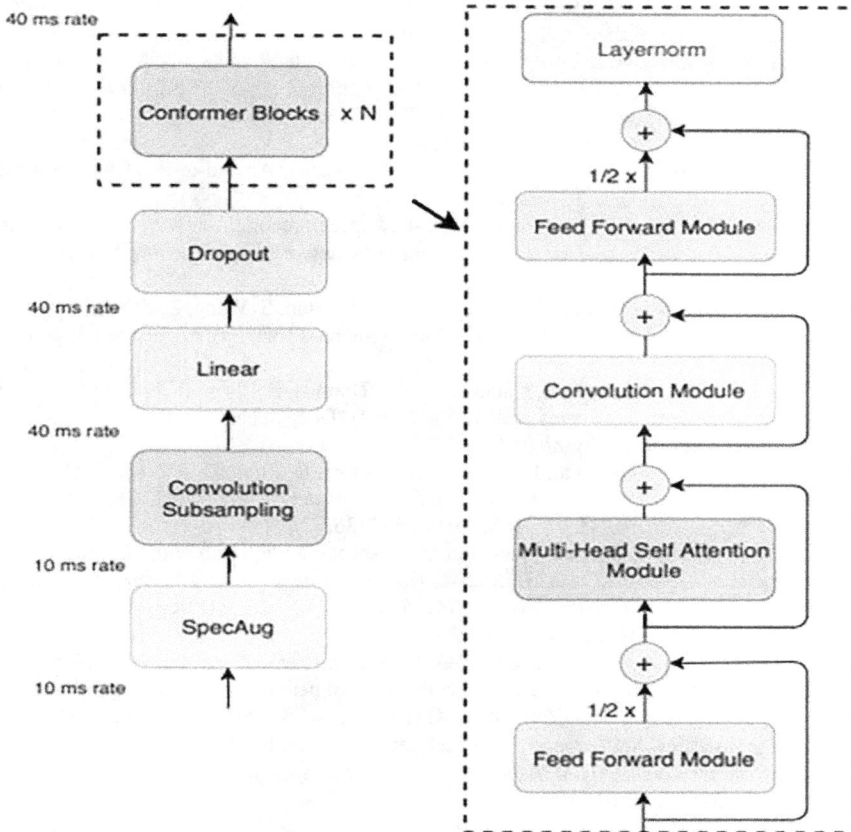

FIGURE 7.19 Conformer Architecture.

7.4 SUMMARY

This chapter explores various models for developing end-to-end ASR pertained models such as CTC, LAS, deep speech, Jasper, QuartzNet, and Wav2vec-based models in detail. Self-supervised models and streaming ASR models were also discussed in detail.

BIBLIOGRAPHY

A. Baevski and Wei-Ning Hsu, Qiantong Xu, Thirunavukkarasu Arun Babu, Jiatao Gu and Michael Auli (2022). data2vec: A general framework for self-supervised learning in speech, vision and language. Proceedings of the 39th International Conference on Machine Learning, Baltimore, Maryland, USA, PMLR 162, 2022. Editors: Kamalika Chaudhuri, Stefanie Jegelka, Le Song, Csaba Szepesvari, Gang Niu, Sivan Sabato.

Alexei Baevski, Henry Zhou, Abdelrahman Mohamed, and Michael Auli. 2020. *Wav2vec 2.0: a framework for self-supervised learning of speech representations*. In Proceedings of the 34th International Conference on Neural Information Processing Systems (NIPS'20).

Curran Associates Inc., Red Hook, NY, USA, Article 1044, 12449–12460. Editors: H. Larochelle, M. Ranzato, R. Hadsell, M.F. Balcan, H. Lin.

W. Chan, N. Jaitly, Q. Le and O. Vinyals. (2016). Listen, attend and spell: A neural network for large vocabulary conversational speech recognition. *2016 IEEE International Conference on Acoustics, Speech and Signal Processing (ICASSP)* (pp. 4960–4964). New York: IEEE. doi: 10.1109/ICASSP.2016.7472621.

R. Collobert, C. Puhrsch and G. Synnaeve. (2016). Wav2Letter: An end-to-end ConvNet-based speech recognition system. *ArXiv, abs/1609.03193.*

A. Graves and N. Jaitly. (2014). Towards end-to-end speech recognition with recurrent neural networks. *Proceedings of the 31st International Conference on Machine Learning.* Editors: Eric P. Xing, Tony Jebara 32(2): 1764–1772.

A. Gulati, J. Qin, C. Chiu, N. Parmar, Y. Zhang, J. Yu, W. Han, S. Wang, Z. Zhang, Y. Wu and R. Pang. (2020). Conformer: Convolution-augmented transformer for speech recognition. *ArXiv, abs/2005.08100.*

A. Y. Hannun, C. Case, J. Casper, B. Catanzaro, G. F. Diamos, E. Elsen, R. J. Prenger, S. Satheesh, S. Sengupta, A. Coates and A. Ng. (2014). Deep speech: Scaling up end-to-end speech recognition. *ArXiv, abs/1412.5567.*

Y. He, T. Sainath, R. Prabhavalkar, I. McGraw, R. Alvarez, D. Zhao, D. Rybach, et al. (2019). *Streaming end-to-end speech recognition for mobile devices* (pp. 6381–6385). Brighton, UK: ICASSP. doi: 10.1109/ICASSP.2019.8682336.

S. Kriman et al. (2020). Quartznet: Deep automatic speech recognition with 1D time-channel separable convolutions. *ICASSP 2020–2020 IEEE International Conference on Acoustics, Speech and Signal Processing (ICASSP)* (pp. 6124–6128). New York: IEEE. doi: 10.1109/ICASSP40776.2020.9053889.

J. Li, V. Lavrukhin, B. Ginsburg, R. Leary, O. Kuchaiev, J. M. Cohen, H. Nguyen and R. T. Gadde. (2019). Jasper: An end-to-end convolutional neural acoustic model. *Interspeech.*

H. Wei-Ning, B. Benjamin, H.T. Yao-Hung, K. Lakhotia, R. Salakhutdinov and A. Mohamed (2021). HuBERT: Self-supervised speech representation learning by masked prediction of hidden units. *IEEE/ACM Transactions on Audio, Speech, and Language Processing* 29: 3451–3460. doi: 10.1109/TASLP.2021.3122291.

8 Computer Vision Basics

LEARNING OUTCOMES

After reading this chapter, you will be able to:

- Understand the CV tasks such as pre-processing, feature extraction, and image classification.
- Understand the basic concepts behind CV and image processing.
- Identify the tools, techniques, and datasets used for image processing and CV applications.

8.1 INTRODUCTION

Computer vision is a technology solution that can assist computers in seeing and fully comprehending image content such as photographs and videos. It is primarily an unsolved problem due to a lack of understanding of physiological vision as well as the complexity of human vision in a dynamic and nearly infinitely different physical world. It is a broad field that can be broadly classified as a subfield of machine learning algorithms, and it can utilize both specialized and general learning techniques. It can appear jumbled as an interdisciplinary field of knowledge, with methods borrowed and recyclable from a variety of widely divergent computer science fields. A handcrafted quantitative research approach may be appropriate for one aspirational problem, but another may necessitate a massive and complicated combination of generalized machine learning algorithms. The aim of digital vision is to understand the information in digital images. Generally, this involves interest and motivation to mimic human vision capability. Recognizing the image content may entail first extracting a description from the picture, which could be an item, a word document, a three-dimensional concept, or something entirely different.

A subfield of computer science is CV the objective of which is to build machines, so that it can process and interpret images and the videos just like the human visual system does.

In general as is evident from Figure 8.1, the eye's job is to transform light into nerve impulses, which the brain then uses to create images of our environment.

In Figure 8.2, CV utilizes machine learning approaches and algorithms in order to recognize, differentiate, and classify objects according to their size or color, as well as to find and decipher patterns in visual data such as images and videos.

8.1.1 FUNDAMENTAL STEPS FOR COMPUTER VISION

There are four major steps involved in CV:

DOI: 10.1201/9781003348689-8

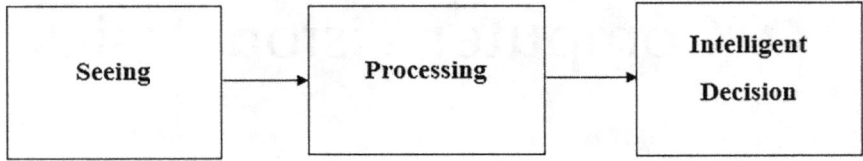

FIGURE 8.1   Human Vision.

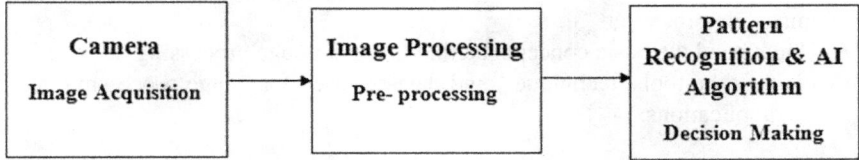

FIGURE 8.2   Computer Vision System.

- Pre-processing
- Segmentation
- Feature extraction
- Classification

In CV, the image analysis system first preprocesses the input image, and after that, segmentation has to be done. Segmentation means the partitioning of an image into connected homogeneous regions. After this, extracting features from the images and features vector takes place. Finally, part of the classification of the image or image recognition is done as shown in Figure 8.3.

8.1.2   Fundamental Steps in Digital Image Processing

8.1.2.1   Image Acquisition

Every system for processing images must start with picture acquisition. Any image captured generally aims to convert a real-world optical image into a range of numerical data that can then be processed by a computer.

8.1.2.2   Image Augmentation

Image augmentation provides fresh data that can be utilized for model training by altering the already-existing data. It can also be defined as the practice of artificially extending the dataset that is accessible for deep learning model training. Image rotation is among the most widely utilized augmentation methods. The data on the image does not change even when it is rotated. Image shifting is one of the method of image enhancement. The orientation of objects in an image can be changed by changing the photos, which gives the model a greater diversity. It ultimately might lead to a more comprehensive model. The image flipping is an extension of image rotation methods. It enables everyone to flip the images in both the up–down and left–right directions.

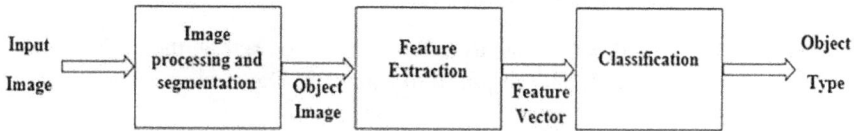

FIGURE 8.3 Typical Block Diagram of a Computer Vision System.

8.1.2.3 Image Enhancement

Image enhancement is the technique of drawing attention to certain aspects of an image while weakening or eliminating any irrelevant details to suit particular requirements.

- Image de-noising: The procedure of wiping out or limiting commotion from the picture is known as a clamor evacuation calculation. The calculations for sound decrease smooth out the whole picture, leaving regions near contrast limits. This decreases or dispenses with the deceivability of commotion. Be that as it may, these procedures might wreck little, low-contrast subtleties.
- Image deblurring: A procedure called picture deblurring recovers a clear latent picture from a blurred image that has been distorted by a camera shake or moving objects. In the domains of image processing and CV, it has drawn a lot of interest.

8.1.2.4 Image Restoration

Picture rebuilding is a point that spotlights on upgrading the part of a photo. Picture rebuilding, in contrast to expansion, is level headed as reclamation strategies are generally subject to science or likelihood models for picture corruption.

8.1.2.5 Color Image Processing

The analysis, manipulation, and interpretation of color-presented visual data constitute color image processing. It can provide a variety of outcomes, from the conversion of black and white image to grayscale to an in-depth examination of the data included in a picture.

8.1.2.6 Image Resizing

Scaling an image is referred to as image resizing. Many applications of image processing and machine learning benefit from scaling. It assists in lowering the number of pixels, which has a number of benefits, like by increasing the amount of input nodes, which in turn raises the complexity, as it can shorten the time required to train a neural network.

8.1.2.7 Multi-resolution Processing

Wavelets are used to express images with differing resolutions. For compression methods and pyramid recognition, images are broken down into smaller areas.

8.1.2.8 Compression

Compression is involved with ways to reduce the image size without degrading its quality as well as the bandwidth required to submit it. Data compression is critical, especially when using the Internet.

8.1.2.9 Morphological Processing

Morphology is a broad category of image-handling strategies, which breaks down images based on structures. Morphological cycles add an organizing component to an info picture to control the resulting picture of a similar size.

8.1.2.10 Image Formation

In any image generation process, mathematical crude and transformations are required for projecting 3D geometric characteristics into 2D features. Image generation is influenced by discrete color and intensity values in addition to geometric elements. It must be aware of the lighting in the environment, camera optics, sensor qualities, and so on.

8.2 IMAGE SEGMENTATION

Image segmentation is a technique used in digital image processing and analysis to divide a picture into different parts or region, usually based on the pixels in the picture. Various types of image segmentation are thresholding techniques, the split and merge techniques, region green techniques, active control, watershed algorithm, and k-means clustering. Image segmentation is pre-processing step of CV systems, and, after image segmentation, it can extract image features for image classification and image recognition.

Mathematically, the segmentation problem partitions image I into regions $R_1, R_2, ..., R_N$. such that

1. $I = R_1 \cup R_2 \cup ... \cup R_N$
2. $R_i \cap R_j = \varnothing , i \neq j$
3. There is a predicate that exists such that P(Ri)= True as well as for adjacent (i, j) P$\left(R_i \cup R_j \right)$ = False ·

8.2.1 STEPS IN IMAGE SEGMENTATION

Image segmentation has six primary types of techniques that are

- Thresholding
- Region growing
- Region split and merge
- Edge based
  - User-stored
  - Active contour
- Topology-based
  - Watershed
- K-means clustering

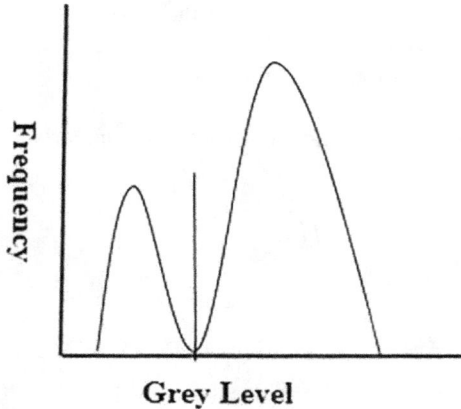

Grey Level

FIGURE 8.4 Threshold Image.

8.2.1.1 Thresholding

Thresholding is a kind of picture division wherein we change an image's pixel creation to work with the examination. Through thresholding, we transform a variety or grayscale pictures into paired pictures.

In thresholding, the image histogram is shown in Figure 8.4. Y axis will be representing frequency, and the x axis will be representing grey level. Based on this model, if $I(x, y)$ is less than the threshold, then the object is a character, and else the background is paper. Assume that a picture, $f(x, y)$, is made out of light items on a dull foundation, and the accompanying figure is the histogram of these pictures. The item can then be extracted by contrasting the pixel value and an edge T.

Advantage is as follows:

- It is a method for making a parallel picture from a grayscale or full-variety picture.

Disadvantage is as follows:

- Nonuniform lighting may cause the histogram to change, making image segmentation impossible with a single global threshold.

8.2.1.2 Region Growing

Region growing is a sequence of region-based segmentation techniques that assemble pixels into wide areas based on predefined grain pixels, growing requirements, and stop conditions. In case the client is expected to choose a point in a specific element implies, suppose in a specific region or a particular feature, corresponding to this color, particular color in a particular region must choose a specific point. The region is used by adding pixels that are comparative in brilliance or variety.

The image and seed point is $S_1, S_2, S_3, \ldots$, so, the difference in intensity concerning the seed point is less than five. The following are the steps involved in pixel segmentation:

- Starting-signal point regions consist of the seed points.
- Add the seed points to the pixels for which the predicate is true.
- Repeat until all the pixels are segmented.

For example, in the case of color images, these comparisons have to be made in terms of RGB values. Suppose

$$I(x, y) = R_1\, G_1\, B_1$$

$$I(x^1, y^1) = R_1^1\, G_1^1\, B_1^1$$

$$\text{Compare} = \sqrt{\left(R_1 - R_1^1\right)^2 + \left(G_1 - G_1^1\right)^2 + \left(B_1 - B_1^1\right)^2}$$

So, like this finding of the similarities between two pixels, the predicate is considered to determine the homogeneity condition at the seed points to the pixels for which the predicate is true, and this procedure has to repeat until all the pixels are segmented out. This concept of region growth is shown in Figure 8.5.

Advantages are as follows:

- It is a simple technique.
- It is adaptive to gradual changes or sound noise.

Disadvantages are as follows:

- Choosing initial seed points.
- The initial seed point may be obtained by human interpretation or intervention.
- Seed point selection from the modes of the histogram.

Original image   Region growing   Segmented Region

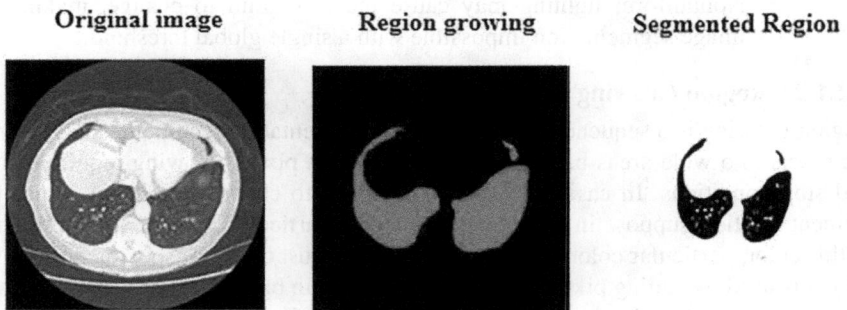

FIGURE 8.5   Difference between Original Image, Region Growing, and Segmented Region.

8.2.1.3 Region-splitting Technique

The basic idea behind region splitting is to divide an image into two different regions that really are coherent inside of themselves. At first, consider the image as the area of interest and examine the area of interest and determine whether all pixels within it satisfy some similarity limit. The opposite approach to local development is local contracting. It is a hierarchical methodology that starts with the suspicion that the overall picture is homogeneous. If this is incorrect, the image is divided into four sub-pictures. This parting strategy is repeated recursively until the image is divided into homogeneous regions.

Figure 8.6 begins with the assumption that the entire image is racially homogenous. If this statement is false, the image is divided into three stages. The first three regions are homogeneous, so there is a need to do the splitting, while the fourth region is heterogeneous. By using a quad tree, the regions are splitted. R is split into four regions, and the regions are $R_1, R_2, R_3, and R_4$. Now $R_1, R_2, and R_3$ are homogeneous' hence, there is no need to do the splitting. R_4 is not homogeneous, and that is why it is divided into four regions. The regions are $R_{41}, R_{42}, R_{43}, and R_{44}$. This procedure repeats and also splits and merges.

Split and merge:

In the event that the district R is non-homogeneous, $P(R) =$ False, then it is split into four districts.

- Combine two contiguous locales R_i and R_j assuming that they are homogeneous. Then $P(R\_iUR\_j) =$ True.
- Stop when no further split or merging is possible.

8.2.1.4 Active Control-based Techniques

Active control is a segmentation technique that separates the important pixels from an image for further analysis and testing using energy forces and limitations. A salient curve

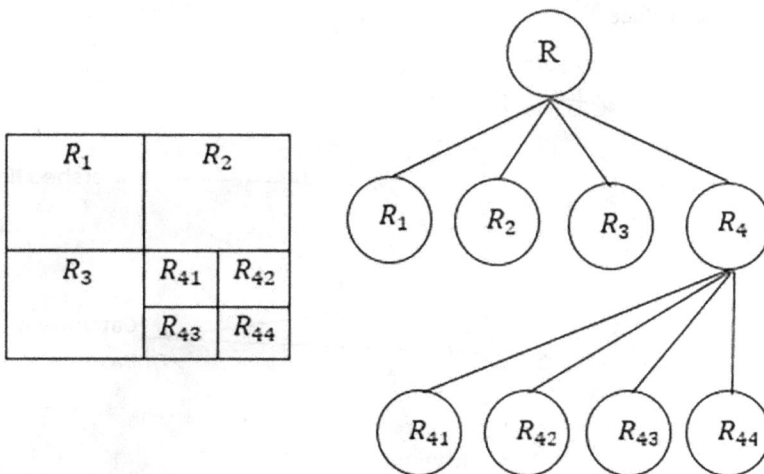

FIGURE 8.6 Region-splitting Techniques.

is generated by minimizing snake energy, initial energy (curve bending and continuity), external energy (Image every [gradient]), and constraint energy (proportion of outside limitations either from more elevated level shape data or client-applied energy).

The following are the steps involved in modeling:

- The shape is characterized in the (x, y) plane of a picture as a parametric bend.

$$V(S) = (X(S), Y(S)) \tag{8.4}$$

- Form is said to have energy (E_{snake}) which is characterized as the amount of the three energy terms.

$$E_{snake} = E_{internal} + E_{external} + E_{constraint}$$

- The energy terms are characterized shrewdly in a manner to such an extent that the last place of the shape will have the least energy.

Advantage is as follows:

- Works well for ambiguous boundary regions.

Disadvantage is as follows:

- The trade-off between noise reduction and detailed boundaries.

8.2.1.5 Watershed Technique

Another popular image segmentation technique, which is the topographic base segmentation technique, is a watershed algorithm. In this case, the image is modeled as a topographic surface.

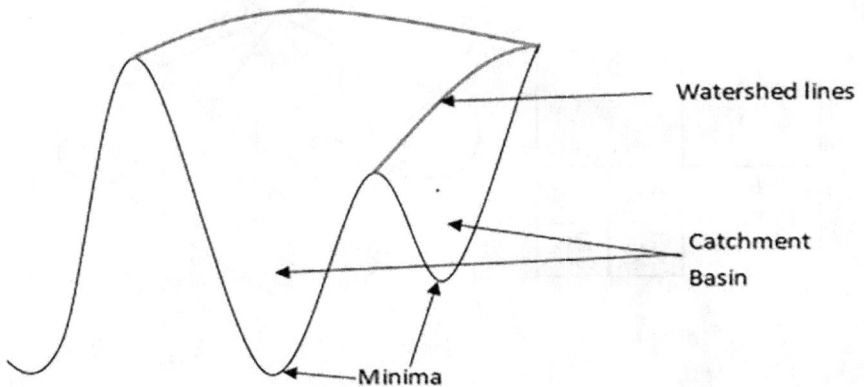

FIGURE 8.7   Watershed Technique.

Figure 8.7 shows that corresponding to this topographic surface, two catchment basins are available, and corresponding to the catchment basin it has the minima; and one minimum corresponds to one catchment basin and another minimum is corresponding to the second catchment basin. Finding the difference between these two catchment basins is said to be known as the red watershed line.

Advantages are as follows:

- Close boundary
- Correct boundary achievable

Disadvantages are as follows:

- Difficulties in correct and efficient implementation of watershed
- Over segmentation problem
- The boundary may not be smooth.

8.2.1.6 K-means Clustering

Image segmentation aims to transform an image representation into something more insightful and understandable. Typically, it is employed to establish borders and locate things as shown in Figure 8.8.

Steps involved in K-means clustering are as follows:

1. Here, K represents the number of clusters.
2. Place the data points into any K clusters at random.
3. Then determine the cluster center.
4. Determine the separation between the data points and the cluster center.
5. Reconfigure the sets of data to the closest clusters on the basis of the separation between each data point and the cluster.
6. Calculate the current cluster center once more.
7. Steps 4, 5, and 6 were repeated as necessary to reach the specified number of iterations or until the data points do not even change the clusters.

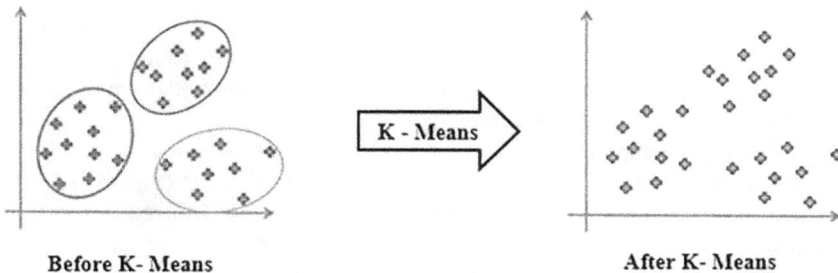

Before K-Means   After K-Means

FIGURE 8.8   K-means clustering.

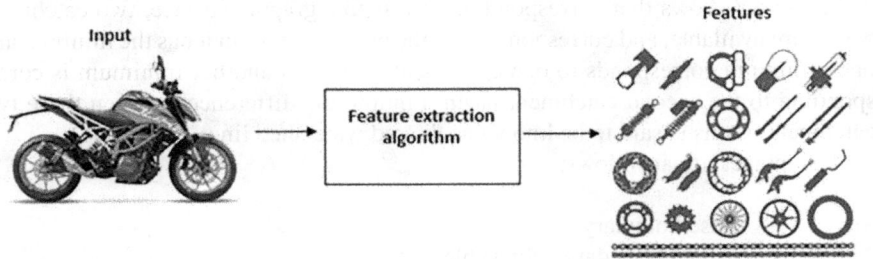

FIGURE 8.9 An Example for Feature Extraction.

8.3 FEATURE EXTRACTION

A layered-decrease method called highlight extraction transforms a lot of crude information into more modest, simpler to-handle gatherings. These tremendous datasets share the characteristic of having numerous factors that request a ton of computational ability to process. The expression "include extraction" alludes to methods for picking as well as consolidating factors into highlights, which essentially reduces how much information that should be handled while appropriately and completely describing the underlying informational index.

An element in AI is a specific measurable property or benchmark for a recognized anomaly. In order to obtain a forecast or classification, features are statistics that you provide into machine learning algorithms. Let's say you need to estimate the cost of a house. The model will recalculate the cost terms based on the input highlights (properties), which may include square footage, occupancy, restrooms, etc. The ability of ML algorithms to anticipate the outcomes is increased by selecting excellent highlights to differentiate the objects.

An element in CV is a quantifiable, object-specific piece of data in an image. It could be a specific shade of color or a specific shape, such as a line, an edge, or a section of an image. An effective element is used to distinguish different items. In ML projects, the learner should turn the raw input (picture) into an element vector so that the training computation can understand the attributes of the item in Figure 8.9. The component extraction method in the graphic is given the basic information picture of a bike. Let's think of the component extract calculation as a black box for the time being, and then we'll turn it back. The element vector, a one-dimensional representation, serves as a representation for the objects.

It is probably manageable without the wheel of one specific bike which has a complete control. When all other criteria are equal, it appears to be a roundabout form for certain examples that identify wheels in all photos in the training dataset. Figure 8.10 depicts the result of detecting a unique bike wheel image among thousands of wheel images. When the component extractor comes across a large number of bike images, it recognizes wheel designs that are common across all cruisers, regardless of where the exit is in this image or even what type of boat they were required for. The good element will assist us in perceiving an article in all of its possible forms. A good element is easily identifiable; simple to understand and apply; and is reliable across measurements, lighting conditions, and evaluation points, which are the qualities it has.

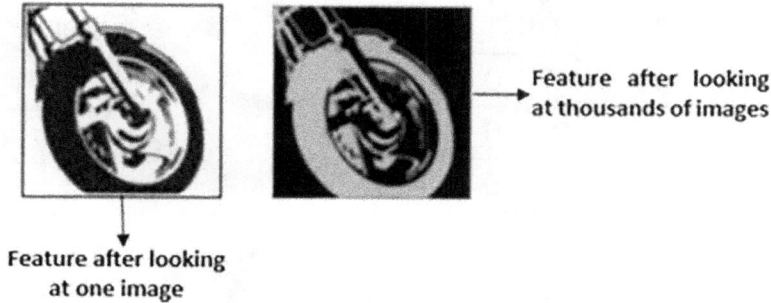

Feature after looking
at thousands of images

Feature after looking
at one image

FIGURE 8.10   Features Need to Detect General Patterns.

8.4   IMAGE CLASSIFICATION

The most common way of assessing a specific class, or mark, with something depicted as a bunch of pieces of information, is known as picture characterization. Image analysis is a subcategory of the clustering algorithm in which a label is assigned to an image sequence.

8.4.1   IMAGE CLASSIFICATION USING CONVOLUTIONAL NEURAL NETWORK (CNN)

Convolutional neural networks are a type of deep, feed forward neural network. CNN comprises neurons with learnable loads and predispositions, very much like brain organizations. CNN has several hidden layers, as well as an output layer. Each neuron receives a large number of i/p, computes a weighted sum, and then runs it through an activation function. A ConvNet's plan resembles the organizational illustration of neurons in the human brain and is spiced up by the relationship of the visual cortex. ConvNet's job is to resize the pictures into an organization that is more straightforward to process without losing highlights that are significant for making precise expectations. Activity of the CNN will likely concentrate on undeniable level highlights like edges from the information image.

The components of CNN are as follows:

- Convolutional layer
- Pooling or down-sampling layer
- Flattening layer
- Fully connected layer

8.4.2   CONVOLUTION LAYER

A convolution layer contains several filters that can perform convolution operations. The base units in this system are filters or kernels. These layers are made up of a number of learnable filters. Convolution is accomplished by computing the dot product of the input matrix and the filter or kernel.

1	0	0	0	0	1
0	1	0	0	1	0
0	0	1	1	0	0
1	0	0	0	1	0
0	1	0	0	1	0
0	0	1	0	1	0

Filter 1

1	-1	-1
-1	1	-1
-1	-1	1

3 x 3

6 x 6 image

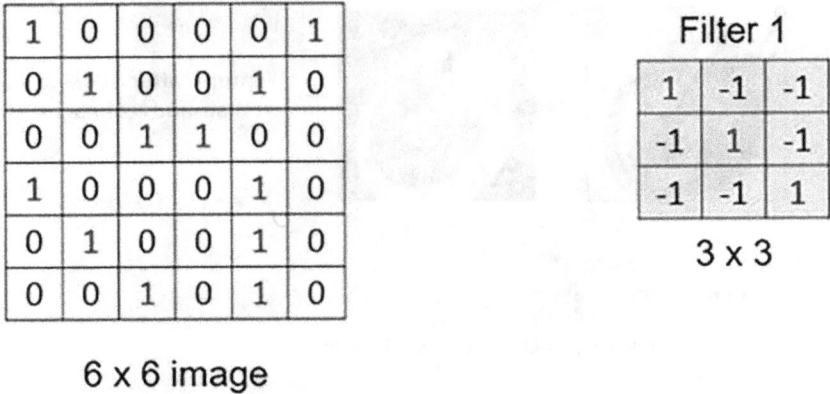

FIGURE 8.11 Understanding the Process of Convolution with an Example.

The kernel iterates over the input vector performing element-by-element matrix multiplication in order to conduct convolution. The element map records the result for each responsive field or the locale where convolution happens. The input image is a 6×6 matrix convolved with a 3×3 filter which produces a featured map. Since the shape of the filter is 3×3, this convolution is called a 3×3 convolution as shown in Figure 8.11. Three parameters used to determine the size of the feature map are as follows:

1. **Depth:** It is the number of filters used for convolution operation. If the convolution is performed on an original image using n filters, then it produces n different feature maps. So the depth of the feature map will be n.
2. **Stride:** The strides are the number of pixels that turn to the input matrix. We shift the filters to 1 pixel at a time when the number of strides is 1. The filters are carried to the next 2 pixels when the number of strides is 2 and so on. They are crucial because they regulate how the filter convolutions happen with the input.
3. **Zero-padding:** Zero-cushioning alludes to the course of evenly adding zeroes to the information lattice. A regularly utilized change permits the size of the contribution to be changed in accordance with the prerequisites. Sometimes, the filter will not fit perfectly with the image. In that case, we need to pad the picture with zeros so that it fits. It is called valid padding, which keeps just the substantial piece of the picture.

8.4.3 POOLING OR DOWN SAMPLING LAYER

The size of the component maps is decreased by pooling layers. Figure 8.12 shows how much organization calculation and the quantity of contributions to be learned. The element map made by a convolution layer's component pooling layer portrays the highlights that are available in a specific region. The most popular filter size is 2×2, which is applied with a stride of 2, and discards 75% of the activations for each depth slice in the input that is down sampled by 2 along both width and height.

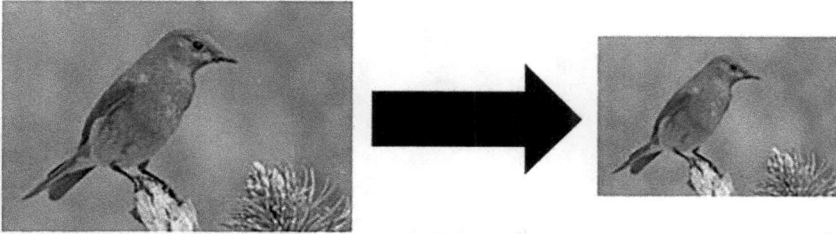

FIGURE 8.12 Example for Down Sampling Image.

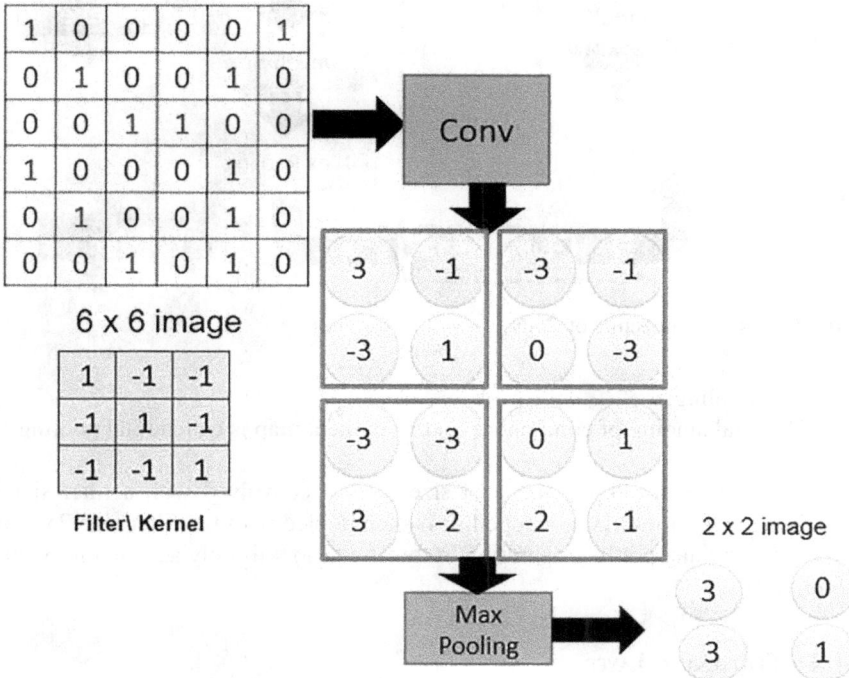

FIGURE 8.13 Pooling Layer.

Different types of pooling are as follows:

- Max-pooling:
  - Returns the best worth from the piece of the image covered by the piece.
- Average pooling:
  - Returns the typical of the huge number of values from the piece of the image covered by the piece.
  - Average pooling works well for straight lines and smaller curves but cannot detect extreme features like sharp edges.

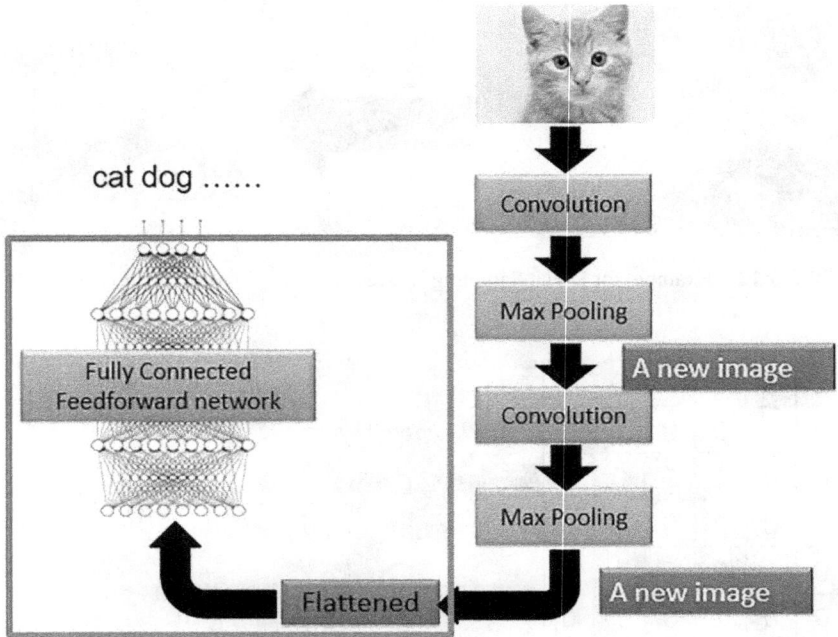

FIGURE 8.14   Architecture of CNN.

- Sum pooling:
- The total amount of components in the element map is called total pooling.

In Figure 8.13, the input image of size 6×6 is convolved with a filter size of 3×3, then the 4×4 matrix is mapped. It is then pooled with the filter size 2×2 and stride 2. Every max pooling operation in this scenario will only accept a maximum of four values.

8.4.4   FLATTENING LAYER

It is time for classification now that the features from the convolution layer have been extracted and the dimension has been reduced by the pooling layer. Multidimensional data cannot be processed by fully connected layers. As a result, before processing, the data should be reduced to a single dimension (flattened) is shown in Figure 8.14.

8.4.5   FULLY CONNECTED LAYER

As in a typical brain organization, neurons in a totally connected layer have a full relationship with all establishments in the past layer. It's a standard Multilayer Perceptron (MLP) with a softmax enactment capability in the result layer. The fully connected layer's purpose is to classify the input image using high-level features from the convolution and pooling layers. Multidimensional data cannot be processed by fully connected layers. As a result, before processing, the data should be reduced to a single dimension (flattened).

8.4.6 ACTIVATION FUNCTION

Complex patterns must be learned and understood by an ANN. To introduce nonlinearity in the convolution layers, the activation layer is applied immediately after each convolution layer.

1. ReLU: Rectified Linear Unit
   - Nowadays ReLU is mostly used as it alleviates the vanishing gradient problem.

$$F(x) = \max(0, x) \tag{6}$$

   - ReLU is very simple to compute because it only requires a comparative evaluation among its input.
   - It has a modified version of 0 and otherwise 1, depending upon whether the input is negative or not.

$$< \mathrm{ReLU}\left(X\right) = \{ _{1,\ for\ x \geq 0}^{0,\ for\ x < 0} \}$$

2. Softmax
   - It is used to predict a multinomial probability for a neural network model.

$$\sigma\left(\vec{Z}\right)_i = \frac{e^{Z_i}}{\sum_{j=1}^{K} e^{z_j}}$$

8.5 TOOLS AND LIBRARIES FOR COMPUTER VISION

8.5.1 OPENCV

A software library for CV is called OpenCV. OpenCV, which was developed to offer a standard infrastructure for image-processing applications, gives users access to more than 2,500 traditional and cutting-edge algorithms.

8.5.1.1 Reading Image in OpenCV Using Python

Use the cv2.imread () function of OpenCV to read a picture in Python. Depending on how many color channels are present in the image, imread() produces either a 2D or 3D matrix. A 2D array will work for a binary or grayscale image. However, a 3D array is required for a colorful image.

8.5.1.2 Writing an Image in OpenCV Using Python

It happens rather frequently in image processing while working with image apps that you need to save the final image or keep intermediate outcomes of image transformations. NumPy n-D array is where images are saved when using OpenCV Python. Use the OpenCV Python library's cv2.imwrite () function to store an image to the file system.

An image can be saved to any storage medium using the cv2.imwrite (). This will save the image in the current working directory using the format that was provided.

8.5.1.3 Displaying Image in OpenCV Using Python

An assortment of Python ties called OpenCV-Python was made to resolve issues with PC vision. A window containing an image is displayed using the cv2.imshow () technique. The window adjusts itself to the size of the image.

8.5.2 MATLAB

Applications involving image, video, and digital signals and AI can all benefit from the programming environment MATLAB. It includes a CV toolkit with numerous features, applications, and algorithms to assist you in creating remedies for CV-related problems.

Architects and researchers can utilize the programming climate MATLAB to review, make, and test frameworks and innovations that can influence the world. MATLAB is a lattice-based language that empowers the most normal articulation of PC math, which is the center of MATLAB.

MATLAB is used for the following:

- Parallel computing:

Utilizes multicore desktops, GPUs, clusters, and clouds to carry out massive computations and parallelizes simulations.

- Interfaces for external languages:

Python, C/C++, Java, and more languages can all be used with MATLAB.

- Platforms that can be integrated:

The Open Neural Network Exchange (ONNX) format, which is supported by MATLAB, allows for the import and export of open-source deep learning frameworks. Direct model import is also possible from PyTorch and Tensor Flow. This enables to integrate the most recent deep learning studies from the community with MATLAB's data labeling tools, data analysis, and GPU program production.

To access the MATLAB desktop from a web browser, we can use Jupyter and MATLAB Integration. The JupyterHub deployment have the Jupyter Notebook

Server for a single user, and many more Jupyter-based provisioning systems that are running locally or in the cloud may all be integrated with MATLAB. By opening it from the Jupyter interface you can work directly in MATLAB without leaving your web browser.

MATLAB can be accessed from another programming environment using MATLAB Engine APIs. The APIs allow MATLAB commands to be executed from inside your programming language even without opening a MATLAB desktop session. There are MATLAB Engine APIs for:

- C/C++
- Fortran
- Java
- Python

Various applications and components, many of which are built in languages like Visual C#,. NET and Visual Basic.NET

8.5.2.1 Toolbox in MATLAB for Computer Vision

A toolbox is just a collection of classes and/or functions. They give you resources, usually for a certain subject (like signal analysis or image processing).

8.5.2.2 Deep Learning Toolbox

A framework for creating and using deep learning models, pre-trained models, and applications is offered by Deep Learning Toolbox Long Short Term Memory (LSTM) networks, and convolutional neural networks (ConvNets, CNNs) can be used for conducting classification images' regression on image, time-series data, and text data. Programmed separation, exceptional preparation circles, and shared loads can be utilized to make network geographies like generative ill-disposed networks (GANs) and Siamese organizations. It might graphically make, assess, and train networks with the assistance of Deep Network Designer programming. It may manage many deep learning projects and also can keep track of various training parameters, examine outcomes, and analyze the code from several experiments using the Experiment Manager tool. Layer activations are visible, and training progress is graphically tracked.

8.5.2.3 Image Processing Toolbox

Image Processing Toolbox provides a comprehensive set of standard calculations and work process tools for image handling, examination, perception, and calculation improvement. To achieve picture division, improvements, commotion evacuation, mathematical changes, and enrollment, advanced learning and regular picture handling strategies can be used. The toolkit supports the handling of two-layered, three-layered, and erratic large images.

Common image-processing procedures can be automated with apps from the Image Processing Toolbox. It can batch-process big datasets, compare picture registration methods, and interactively segment image data and it may explore photos, 3D volumes, and movies using visualization tools and apps, alter contrast, produce histograms, and edit regions of interest (ROIs).

8.6 APPLICATIONS OF COMPUTER VISION

8.6.1 OBJECT DETECTION

Object detection is a computer-based technique that is connected with PC vision and picture handling for recognizing an article in a picture or video. The difference between classification, localization, and detecting is given next.

Various image-processing techniques are shown in Figure 8.15. In classification, given a cat in an image, the objective is to determine whether there is a cat that corresponds to that image. In localization, in addition to saying that there is a cat, it also has a bounding box around the cat to indicate where in the image the cat is located. Object detection takes it a step further by detecting and localizing all possible occurrences of the object in your set of classes in a given space. In a localization task, there is only one object which localizes, whereas, in objection detection, there could be multiple objects detected and multiple instances of the same object. It could have a cat and a dog or both, all of these possibilities. The objective here is to recognize each of these objects as well as localize them using these kinds of bounding boxes. Finally, there is the task of segmentation, where the job is to label each pixel as belonging to one of, say, c classes, depending on the number of classes.

8.6.2 FACE RECOGNITION

Face recognition is involved by facial acknowledgment innovation in iPhone and high-level security frameworks to perceive a face. It must be fit for perceiving the distinctive attributes of your face to forestall unapproved admittance to the cell phone or PC.

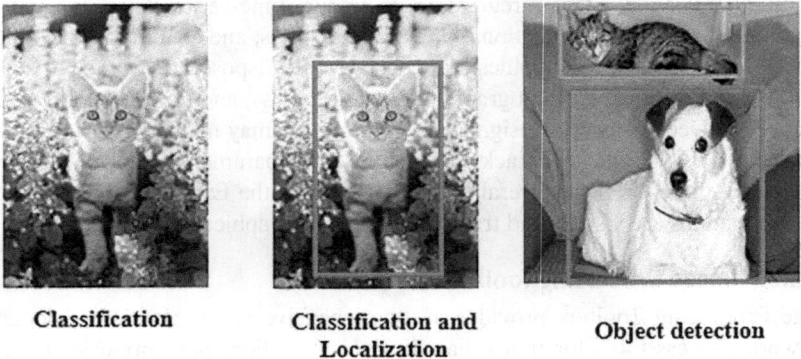

| Classification | Classification and Localization | Object detection |

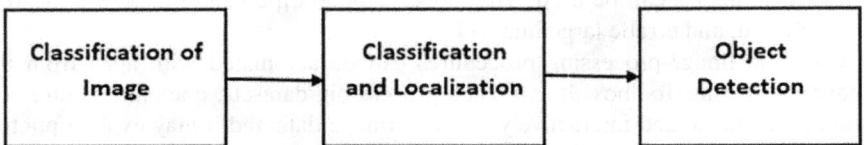

FIGURE 8.15   Steps Involved in Object Detection.

In Figure 8.16, the image data is given as input. The data is pre-processed by using various pre-processing techniques to prepare the data for face recognition. Then the feature extraction takes place, and the process of extracting facial features involves locating the most recognizable facial features such as the eyes, nose, and mouth in photographs of people's faces, and the final part includes training a part of data and test with another set of data.

8.6.3 NUMBER PLATE IDENTIFICATION

Many traffic signals and cameras use tag acknowledgment to accuse punishments and for emergency needs. A traffic framework can perceive a vehicle and get proprietorship data utilizing number plate distinguishing proof innovations. It isolates a number plate and its information from the other, in its vision through picture characterization. This innovation has altogether improved the process of getting fined for states.

The image of the number plate is captured by using camera. Then the number plate of the vehicle is identified and detected. The character segmentation locates the alpha numeric characters which are then translated into alpha numeric text using OCR techniques, and the characters are recognized using ANN as shown in Figure 8.17.

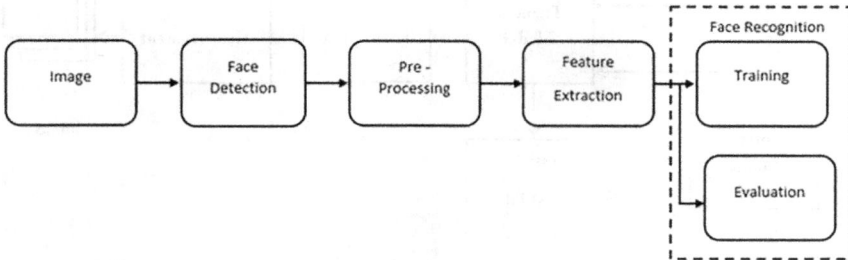

FIGURE 8.16 The Architecture of the Face Recognition System.

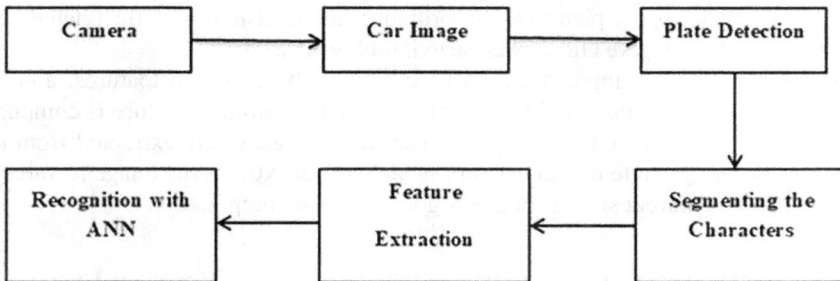

FIGURE 8.17 The Block Diagram of Number Plate Detection System.

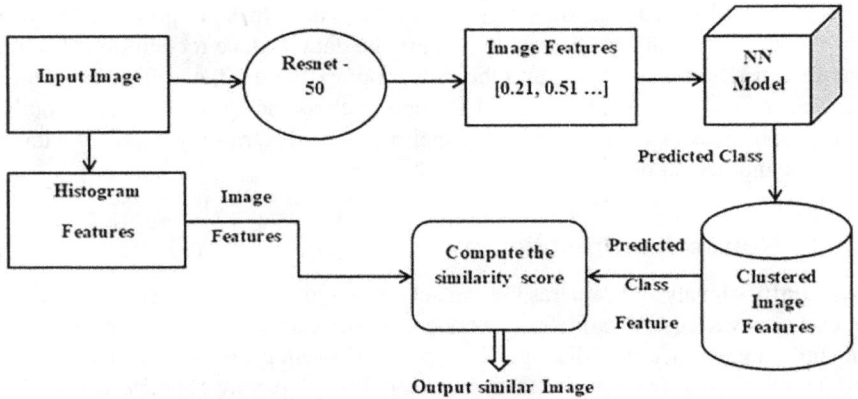

FIGURE 8.18   Architecture for Image-based Search.

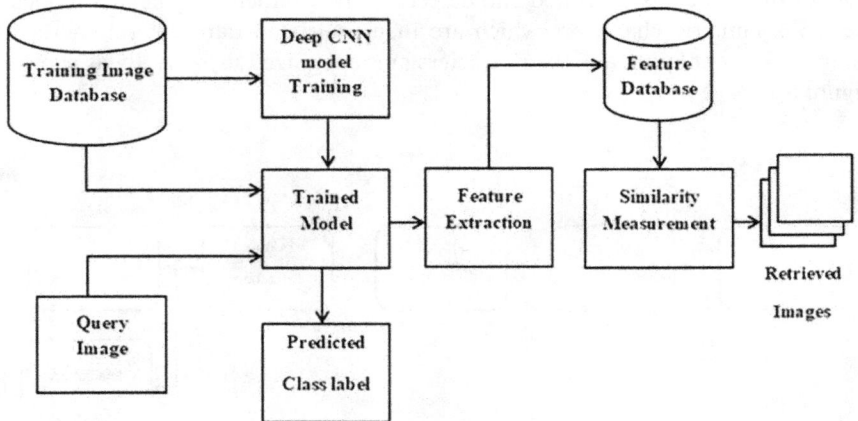

FIGURE 8.19   The Architecture of the Medical Image Retrieval System.

8.6.4   IMAGE-BASED SEARCH

Google and other picture-based web crawlers utilize picture division frameworks to decide the articles in the picture and coordinate their decisions with the related photographs that find to give clients web search tool results.

In Figure 8.18, the input image is processed with histogram features, and the image features are computed for similarity score. The similarity score is computed with the help of predicted class feature. The image features are extracted from the input image and given to neural network model, which extracts the image features as clusters, and the output similar image is got as the final output.

8.6.5   MEDICAL IMAGING

Picture division is utilized in the clinical business to find and recognize growth cells, evaluate tissue volumes, direct virtual careful reenactments, and do in view of the buried route. Picture division has various restorative applications. It supports the recognizable proof of regions impacted and the preparation of suitable consideration.

In Figure 8.19, the training image database is fed to a deep CNN model for training, and then the output is fed to a trained model which also gets a query image as input. The trained model is further processed for feature extraction, which is then fed to a feature database for similarity measurement. Retrieved image is got as one output, and the predicted class label is got as an output from the trained model.

8.7 SUMMARY

In this chapter, the digital image processing and fundamental steps of image processing and various image segmentation techniques are all explained briefly. Also, this chapter has explained the feature extraction techniques and components of CNN. The image classification model was also included. It highlights the tools and libraries for components like OpenCV and MATLAB. Finally, the chapter is concluded with a brief explanation of the applications of CV in real-time. Further, these applications are implemented as case studies in Chapter 10.

BIBLIOGRAPHY

Bayoudh, Khaled, Raja Knani, Fayçal Hamdaoui, and Abdellatif Mtibaa . A survey on deep multimodal learning for computer vision: Advances, trends, applications, and datasets. *The Visual Computer* (2022): 38(8): 2939–2970.

Chowdhary, Chiranji Lal, G. Thippa Reddy, and B. D. Parameshachari. *Computer Vision and Recognition Systems: Research Innovations and Trends*. London: CRC Press (2022).

Guo, Meng-Hao, Tian-Xing Xu, Jiang-Jiang Liu, Zheng-Ning Liu, Peng-Tao Jiang, Tai-Jiang Mu, Song-Hai Zhang, Ralph R. Martin, Ming-Ming Cheng, and Shi-Min Hu. Attention mechanisms in computer vision: A survey. *Computational Visual Media* (2022): 1–38.

Hammoudeh, Mohammad Ali A., Mohammad Alsaykhan, Ryan Alsalameh, and Nahs Althwaibi. Computer vision: A review of detecting objects in videos--challenges and techniques. *International Journal of Online & Biomedical Engineering* (2022): 18(1).

Rani, Shilpa, Kamlesh Lakhwani, and Sandeep Kumar. Three-dimensional objects recognition & pattern recognition technique; related challenges: A review. *Multimedia Tools and Applications* (2022): 1–44.

Tong, Kang, and Yiquan Wu. Deep learning-based detection from the perspective of small or tiny objects: A survey. *Image and Vision Computing* (2022): 104471.

Yang, Xi, Jie Yan, Wen Wang, Shaoyi Li, Bo Hu, and Jian Lin. Brain-inspired models for visual object recognition: An overview. *Artificial Intelligence Review* (2022): 1–49.

Zeng, Kai, Qian Ma, Jia Wen Wu, Zhe Chen, Tao Shen, and Chenggang Yan. FPGA-based accelerator for object detection: A comprehensive survey. *The Journal of Supercomputing* (2022): 1–41.

9 Deep Learning Models for Computer Vision

LEARNING OUTCOMES

After reading this chapter, you will be able to:

- Understand the concepts for applying deep learning to computer vision.
- Understand architectures of pre-trained models in computer vision.

9.1 DEEP LEARNING FOR COMPUTER VISION

In the past, building a database of photos was required for a computer to recognize images. Then, in order for an application to compare and make sense of the visual input, these photos required to be tagged with specific information. By enabling programmers to create features or little apps that could recognize patterns in photos, machine learning enhanced this laborious procedure. A learning algorithm might then be used by these applications to categorize photographs and recognize objects. Machine learning still depends on organized, labeled data, though less so than it did for earlier human procedures. Even unstructured data could be used to arrange the data after some pre-processing. The advantage of deep learning over machine learning is to do all the pre-processing and feature extraction process automatically. It can automate feature extraction from unstructured data, enabling algorithms to learn from the data with little assistance from experts.

The issues of object detection and classification are effectively enough to recognize known items in untrained data, as many CV algorithms require training on thousands of data points. These are issues that frequently occur as a result of bad or insufficient training data, which prevents models from properly generalizing beyond the training data or identifying objects in novel imagery. The famed ImageNet dataset, which included millions of images arranged into countless categories, became accessible in 2010.

9.2 PRE-TRAINED ARCHITECTURES FOR COMPUTER VISION

9.2.1 LeNet

LeCun et al. proposed the convolutional neural network topology known as LeNet in 1998. LeNet-5, simplistic CNN, and word "LeNet" are used to refer to it. Because its artificial neurons may respond to some of the surrounding cells in the coverage area, CNN, a particular feed forward neural network, is particularly effective at processing large-scale images. LeNet is made up of two primary components: a dense

DOI: 10.1201/9781003348689-9

layer with three fully associated layers and a neural generator with convolution layer. A summary of the architecture is shown in Figure 9.1. LeNet is made up of two primary components: a dense network with three fully associated layers and a neural generator with a convolution layer. These layers commonly enhance the quantity of channels by mapping spatially organized inputs to numerous two-dimensional feature extraction. The output channels of the first convolutional layer are 6, and those of the second are 16. Each of the 22 aggregating processes (stride 2) lowers complexity by a four-fold increase through geographical down sampling. The batch size, the quantity of channels, the height, and the breadth all have an impact on the output form of the multilayer block. The three completely linked layers that make up LeNet's dense block each has 120, 84, and 10 outlets. The 10-dimensional activation function accurately represents the number of possible output layers, given the ongoing categorization.

The structural information contained in images may be effectively utilized by CNN. Few parameters are used in the convolutional layer, which is also a result of its primary features, shared weights, and local connections. The model in this architecture lacks the ability to do image classification, and it will have overfitting problems. AlexNet architecture was later designed to address these problems and perform classification task.

9.2.2 ALEXNET

In order to classify images, AlexNet is a deep-CNN model that was first introduced in ILSVRC-2012, a contest created by Alex Krizhevsky, Ilya Sutskever, and Geoffrey E. Hinton in 2012. Its top most of five test error rate was 15.3%, whereas the phase-2 entry's was 26.2%. Although many more network topologies with many more layers have subsequently emerged, eight layers make up AlexNet. The final three layers are completely connected, and the first five are convolutional. As seen in Figure 9.2, there are also other "layers" in between called pooling and activation.

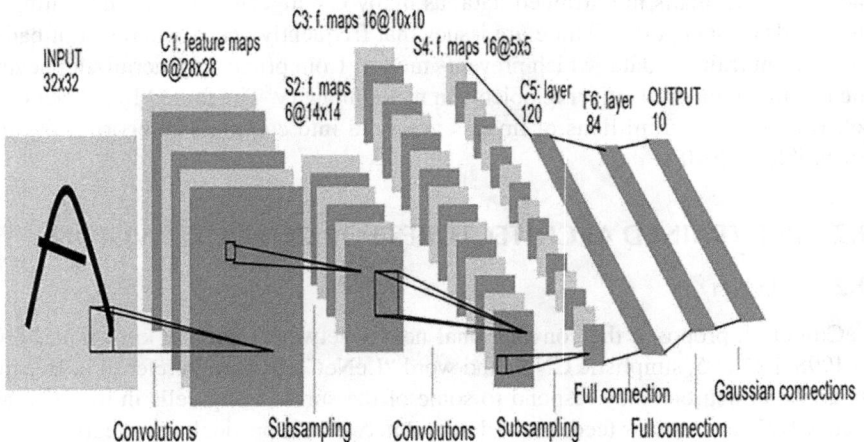

FIGURE 9.1 LeNet Architecture.

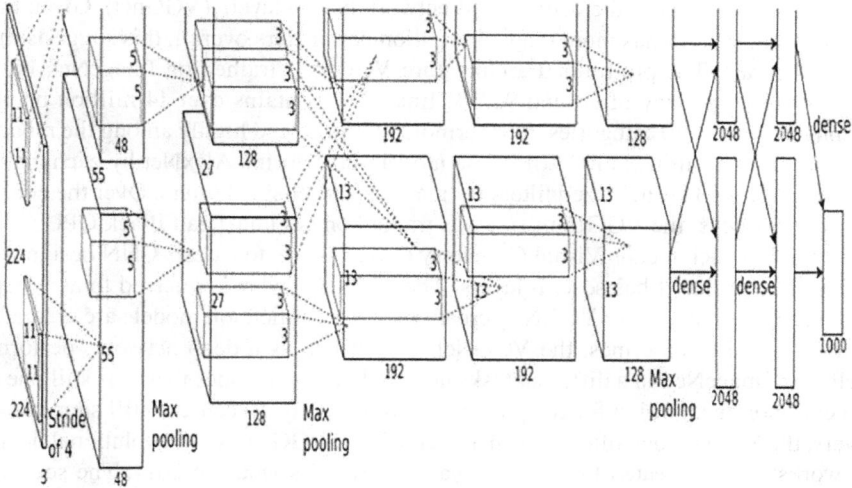

FIGURE 9.2 AlexNet Architecture.

The order of layers of AlexNet is shown in the Figure 9.2. The problem set is split into two sections, with one section running on GPU 1 and the other on GPU 2. Low communication overhead makes it easier to attain overall good performance. The AlexNet results demonstrate that a large, deep convolutional neural network can set new records on a highly difficult set of problems. CNN views AlexNet as a turning point in the classification of wizards. The direct image input to the classification model is AlexNet's distinctive benefit. The supervised learning project AlexNet produced excellent results. Choosing techniques that improved the performance of the network, such as dropout and data augmentation, was also crucial. ConvNets' breakthrough implementations by AlexNet, like ReLU and dropout, are still in use today. Low classification errors without overfitting are difficult to achieve. The architecture's shortcoming is its inability to handle complex applications for high-resolution images. In order to address these shortcomings in AlexNet, the standard consensus network VGG (Visual Geometry Group) was developed to surpass all previous performance thresholds in image recognition tasks. Deep CNNs might be trained significantly more quickly utilizing ReLU nonlinearity than they could using saturating activation functions like tanh or sigmoid, according to AlexNet. Using CIFAR-10 data, it flattens each example in the mini batch before passing through convolutional block and dense block to obtain the outputs.

9.2.3 VGG

The previously discussed 16-layer VGGNet-16 can identify images into 100 separate object categories, such as keyboard, animals, pencil, mouse, and many more. Correspondingly, images with a resolution of 224 × 224 are backed by the model. The core concept of the VGG19 model, also known as VGGNet-19, is similar to that of the VGG16 model, with the exception that it holds 19 layers. The name VGG comprises the number 16 because the deep neural network has 16 layers (VGGnet). Given that the VGG16 network has more than 138 million parameters overall, this suggests that it is quite vast. The pre trained architecture VGG16 is trained on ImageNet, has a top of 5 test accuracy of around 92.7%. ImageNet contains over 14 million photographs in over 1,000 categories. Furthermore, it was placed highly among the models that participated in ILSVRC-2014. The model outperforms AlexNet by exchanging a number of 3×3 kernel-sized filters for massive kernel-sized filters. Over the course of several weeks, the VGG16 model was trained on NVidia Titan Black GPUs. The term VGG, which means Visual Geometry Group, refers to a deep-CNN design that is common in that it has several layers. The count of layers is referred to as "deep" and VGG-16 or 19. Using VGGNet, creative item identification models are created.

In contrast to baselines, the VGGNet, established as a deep network, performs well than ImageNet on a different tasks and on datasets. Besides that, it is still one of the common recognition for Images architectures today. Because it still supports 16 layers, the VGG model, often known as VGGNet or VGG16, is a convolutional neural networks model, created by A. Simonyan and K. Zisserman in 2014. The scenario is explained in Figure 9.3. Even by yesterday's high standards, the model presented by these researchers is a large network and can be found in the publication "Very Convolution Neural Networks for Huge Image Recognition." VGGNet's design is more appealing since it is straightforward.

There are approximately 64 filters that can be accessed, and those numbers can go up to about 128 and eventually to about 256 filters. In the end, 512 filters may be used. For each step or each stack, the number of filters that can be used on a convolution layer doubles. The VGG16 was developed under the direction of this essential notion. Disadvantages of the VGG16 are that it is huge and takes more time to train its parameters. The model is larger than 533 MB due to the model's depth and the number of entirely connected layers. This prolongs the process of implementing a VGG network. The model is utilized for many classification challenges, while smaller network topologies like Google Net and Squeeze Net are also widely used. In any case, the VGGNet is an excellent building block for instructional purposes because it is so easy to implement. The model's inability to be used for deep networks as the network's depth increases, its propensity for the vanishing gradients problem, and the enormous amount of time and money needed for calculation due to the large number of parameters are just a few of the significant disadvantages of this architecture. This issue is then addressed and Inception architecture is introduced to improve speed and accuracy while keeping high performance.

9.2.4 Inception

The improved Inception V3 replaced the basic model Inception V1, which was published as Google Net in 2014. As the name implies, it was formed by a Google team. Data overfitting occurred when a model employed numerous deep layers of convolutions. The Inception models will have parallel layers rather than deep layers, resulting in wider than deep models. The Inception model consists of several Inception modules. Four parallel layers make up the basic module of the Inception V1 model: 1×1 convolution, 3×3 pooling layer, 5×5 batch normalization, and 3×3 pooling

ConvNet Configuration					
A	A-LRN	B	C	D	E
11 weight layers	11 weight layers	13 weight layers	16 weight layers	16 weight layers	19 weight layers
input (224 × 224 RGB image)					
conv3-64	conv3-64 LRN	conv3-64 conv3-64	conv3-64 conv3-64	conv3-64 conv3-64	conv3-64 conv3-64
maxpool					
conv3-128	conv3-128	conv3-128 conv3-128	conv3-128 conv3-128	conv3-128 conv3-128	conv3-128 conv3-128
maxpool					
conv3-256 conv3-256	conv3-256 conv3-256	conv3-256 conv3-256	conv3-256 conv3-256 conv1-256	conv3-256 conv3-256 conv3-256	conv3-256 conv3-256 conv3-256 conv3-256
maxpool					
conv3-512 conv3-512	conv3-512 conv3-512	conv3-512 conv3-512	conv3-512 conv3-512 conv1-512	conv3-512 conv3-512 conv3-512	conv3-512 conv3-512 conv3-512 conv3-512
maxpool					
conv3-512 conv3-512	conv3-512 conv3-512	conv3-512 conv3-512	conv3-512 conv3-512 conv1-512	conv3-512 conv3-512 conv3-512	conv3-512 conv3-512 conv3-512 conv3-512
maxpool					
FC-4096					
FC-4096					
FC-1000					
soft-max					

FIGURE 9.3 VGG Architecture.

layers. By applying a filter to each image and its immediate neighbors throughout the whole image, convolution layers change the image. Pooling is the process used to reduce the dimensionality of the feature map. There are other forms of pooling, but average and maximum pooling are the most popular. One of the main benefits of the Inception model was the extensive dimension reduction. In order to improve the model further, the larger convolutions were divided into smaller convolutions. Fundamental module for the conceptualization V1 module serves as an illustration as shown in Figure 9.4. It includes 55 convolutional layers, which, as already mentioned, required expensive computing. Thus, to reduce computational cost, the 55 convolutional layers were swapped out for two 33 convolutional layers. The fewer parameters also resulted in lower computing costs. The architecture is constrained by a representational problem that reduces the feature space of the layer below, which can end in the loss of a significant amount of details. A program that can detect photos without losing any information was developed using R-CNN.

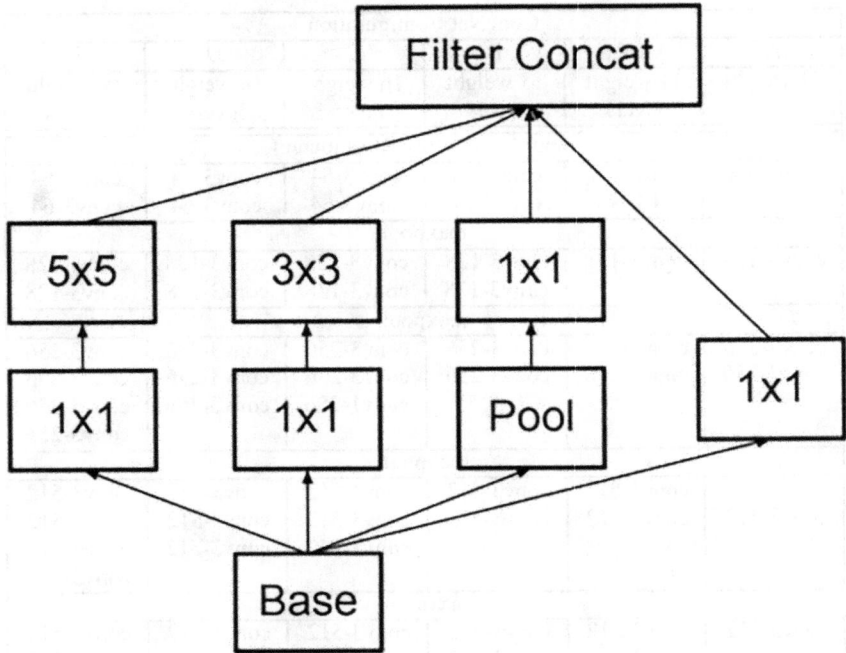

FIGURE 9.4   Inception Architecture.

9.2.5   R-CNN

The region-based convolution neural network, or R-CNN for short, was developed in 2014 by a group of academics at UC Berkeley and is capable of identifying 80 different types of objects in pictures. R-primary CNN's contribution to object identification is limited to feature extraction using a CNN, as opposed to the full object detection procedure shown in the previous image. Figure 9.5 (R-CNN architecture) depicts how the R-CNN model operates. The first module produces recommendations for 2,000 regions by employing the Selective Search methodology. The second module shrinks each region proposal to a predetermined predefined size before extracting feature vector length 4,096. A pre-trained SVM system selects the classification model with the dark background or one of the object classes in module three. The comprehensive dimension reduction was one of the main advantages of the Inception model.

The R-CNN designs also have the following drawbacks. In order to improve the model further, the larger convolutions were divided into smaller convolutions. Consider the basic module of the conceptualization V1 module, which is depicted in Figure 9.5. It includes 55 convolutional layers, which, as already mentioned, required expensive computing. Thus, in order to reduce the computational cost, the 55 convolutional layers were swapped out for two 33 convolutional layers. The fewer parameters also resulted in lower computing costs. To get beyond R-CNN design restrictions, the Fast R-CNN model is recommended as an enhancement to the R-CNN model.

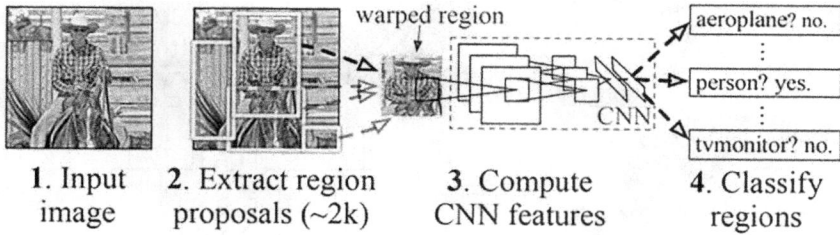

warped region

aeroplane? no.
⋮
person? yes.
⋮
tvmonitor? no.

CNN

1. Input
image

2. Extract region
proposals (~2k)

3. Compute
CNN features

4. Classify
regions

FIGURE 9.5 R-CNN Architecture.

9.2.6 Fast R-CNN

A primary developer of the object detector Fast R-CNN is Facebook AI researcher and former Microsoft researcher Ross Girshick in 2015 [7]. Fast R-CNN can solve several R-CNN issues, as its name implies. The new layer, dubbed ROI Pooling, recovers feature vectors of the same length from each proposal (ROI). In contrast to R-CNN, which has three stages, Fast R-CNN only has one phase (region proposal generation, feature extraction, and classification using SVM). Because it conducts calculations (like those for convolutional layers) only once and then divides them among all of the proposals, R-CNN is quicker (such as ROIs). Fast R-CNN is thus faster than R-CNN, which is achieved by incorporating the latest ROI pooling layer. In contrast to R-CNN, which requires hundreds of gigabytes of disc space, Fast R-CNN will not collect the features extracted. It is a bit less accurate than R-CNN. Figure 9.6 depicts the Fast R-overall CNN design. R-three CNN has three steps, whereas the model has one. It only requires a photo as input and returns the bounding boxes and classifier of the detected items. The feature map from the preceding convolutional layer is fed into a ROI pooling layer. The rationale is to retrieve a length that is fixed with the vector of each region suggestion. The ROI layer operates by dividing each region proposal into cells in a grid. To return a single value, each cell in the grid obtains the maximum pooling operation.

The vector map for feature extraction is made up of values in the cells. If grid size is said to be 22, then feature vector length is 4. Following that, the ROI pooling-extracted feature vector is passed to a few fully connected layers. Each region proposal made using R-CNN is fed to the model separately from the others. Fast R-CNN's search region proposal generation approach consumes the majority of its processing time during detection. Therefore, Faster R- CNN's focus was on the architecture's bottleneck.

9.2.7 Faster R-CNN

Faster R-CNN is an improvement on Fast R-CNN. Faster R-CNN outperforms Fast R-CNN due to the region proposal network (RPN), which was developed in 2015 by Shaoqing Ren, Kaiming He, Ross Girshick, and Jian Sun. A fully convolutional network, the region proposal network (RPN), generates proposals with varying scales and aspect ratios. Anchor boxes are not the same as photo pyramids or filter

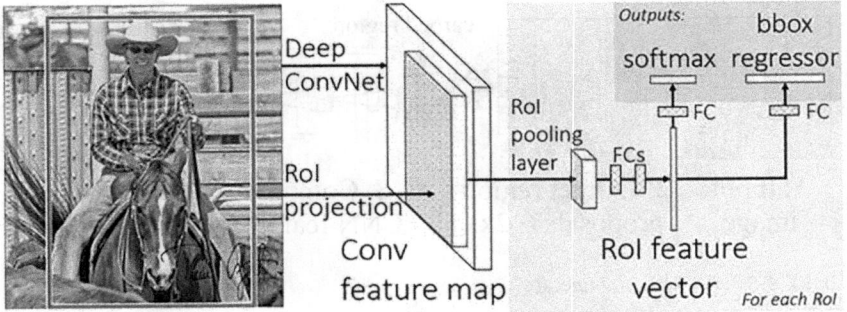

FIGURE 9.6　Fast R-CNN Architecture.

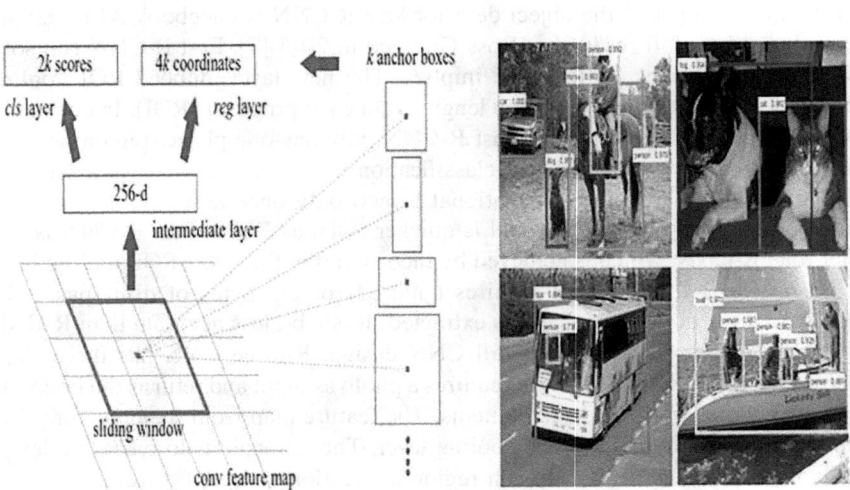

FIGURE 9.7　Faster R-CNN Architecture.

pyramids. An anchor box is a references box with a certain scale and flexural modulus. The same region can have multiple sizes and dimensions if there are many reference anchor boxes. The mapping of each region to a unique reference anchor box that results allows for the recognition of objects with varying scales and aspect ratios. The convolutional calculations used by RPN and Fast R-CNN are the same. As a result, the computation time is cut in half. Figure 9.7 depicts the layout of the Faster R-architectural CNN. The RPN module is in charge of creating region suggestions. Attention is implemented using neural networks, which instruct the Fast R-CNN module for the detection on the place to look upon the object within the given picture.

Regional proposals are created by the RPN. Using the ROI pooling layer, each area suggested in the image is used to extract a fixed-length feature vector. The output feature vectors are then classified using the Fast R-CNN. Along with their bounding boxes, the class scores of the discovered objects are returned. The selective search technique is used by the R-CNN and Fast R-CNN algorithms to produce region recommendations. Every suggestion is routed to a previously trained CNN. All of the 256-bit mini-batch anchors from a single image are used to train the RPN, which is a disadvantage of faster R-CNN. Because all samples from a single image can be associated, the network may take some time to complete (i.e., their features are similar). Faster-RCNN has some flaws, such as the inability to perform real-time detection. Before classification, the procedure for obtaining region boxes entails a number of computations. This constraint facilitated the development of another novel approach, the Mask-RCNN.

9.2.8 MASK R-CNN

The Mask R-CNN model, developed by Kaiming He, Georgia Gkioxari, Piotr Dollar, and Ross Girshick in 2017, extends Faster R-CNN by including an additional branch that returns a mask for each recognized item. Faster R-CNN is a region-based convolutional neural network that creates bounding boxes and a confidence score for the class label of each object. Mask R-CNN, also known as Mask RCNN, is a cutting-edge categorization model based on Faster R-CNN. Mask R-CNN is the most advanced CNN for image segmentation. Understanding how Mask R-CNN works necessitates an understanding of image segmentation. Image segmentation is a CV technique for dividing a digital image into distinct regions (sets of pixels, also known as image objects). This segmentation includes both borders and objects (lines, curves, etc.). Mask R-CNN supports two basic types of image segmentation: semantic and instance segmentation. Semantic segmentation categorizes each pixel without distinguishing between various instances of the same object. To put it another way, semantic segmentation attempts to recognize and group similar items into a single category at the pixel level. Background segmentation, also known as semantic segmentation, is the process of separating the subjects of an image from its background. The method of accurately identifying and finely segmenting each object in an image is known as instance segmentation, also known as instance recognition. As a result, it combines object detection, localization, and classification. In other words, this type of segmentation takes it a step further by separating each item designated as a comparable instance. Despite the fact that all objects in a segmentation scenario are individuals, each individual is given individual attention throughout the procedure. Semantic segmentation is also known as foreground segmentation because it highlights the subjects of the image rather than the background.

Mask R-CNN was created using faster R-CNN, as shown in Figure 9.8. For each candidate item, Faster R-CNN outputs a class label and a bounding box offset, whereas Mask R-CNN adds a third branch that outputs the object mask. Since it differs from the class and box outputs, it requires the retrieval of a much more precise spatial arrangement of an object. CNN's Mask R-CNN extends its functionality by trying to introduce a branch for predicting an object mask (region of interest) as well as a branch for bounding box detection. Fast/Faster R-CNN has an important feature that Mask R-CNN does not have—pixel-to-pixel alignment. Mask R-CNN uses the same two-step procedure, beginning with the same first stage (which is RPN). Mask R-CNN must generate a binary mask for each ROI in relation to class and box offset predictions in the second stage. The majority of modern systems, on the other hand, rely on mask predictions for classification. Furthermore, the Faster

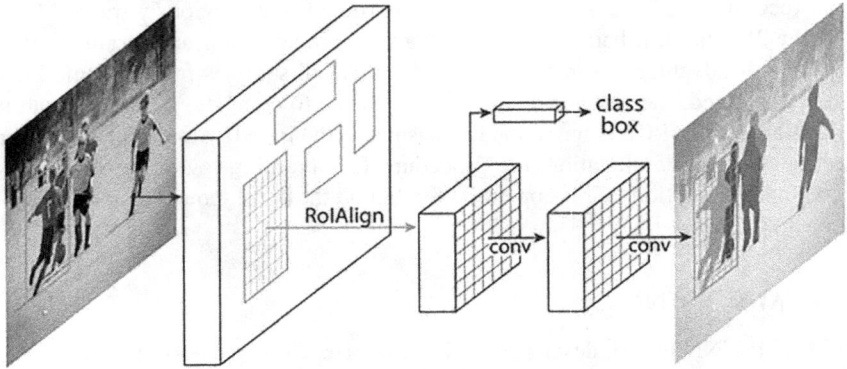

FIGURE 9.8   Mask R-CNN Architecture.

R-CNN framework, which provides a variety of customized architecture designs, simplifies the creation and training of Mask R-CNN, and the low computing cost of Mask R-low CNN enables rapid system development and testing. A number of constraints apply to Mask R-CNN. Because it can only analyze static photographs, it cannot evaluate temporal information about the object of interest, such as active hand gestures. Second Mask R-CNN is typically unable to distinguish between hands and other low-resolution motion-blurred objects. To address this issue, YOLO was created to improve detection performance directly through image training.

9.2.9   YOLO

YOLO is a method that provides real-time object detection using neural networks developed by Joseph Redmon, Santosh Divvala, Ross Girshick, and Ali Farhadi in 2016. The efficiency and speed of this algorithm account for its popularity. It has been used in many different contexts to distinguish between animals, people, parking meters, and traffic signals. YOLO is an abbreviation for You Only Look Once. This algorithm recognizes and locates multiple elements in a photograph (in real-time). The class probabilities of the discovered photographs are provided by the object identification procedure in YOLO, which is carried out as a regression problem. To recognize objects quickly, the YOLO approach employs CNNs. To detect objects, as implied by the name and depicted in Figure 9.9, the method only requires one forward propagation through a neural network. This demonstrates that the image is subjected to a specific algorithm for prediction. CNN is used to forecast multiple bounding boxes and class probabilities at the same time. There are several versions of the YOLO algorithm. Tiny YOLO and YOLOv3 are two of the more well-known versions. The YOLO algorithm is essential for the reasons listed here:

- Speed: It can anticipate objects in real time, and this method speeds up detection.
- Accuracy: In the background, the YOLO prediction method yields precise results with low errors. Because of its superior learning capabilities, the algorithm can recognize and apply object portrayal for object detection.

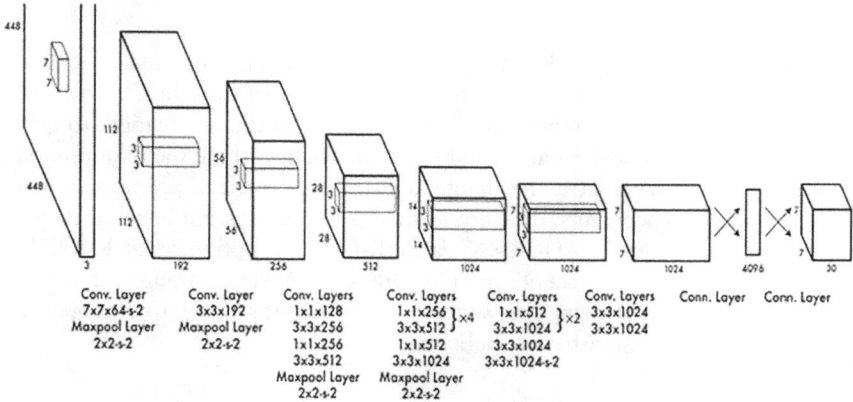

FIGURE 9.9 YOLO Architecture.

The techniques listed here will work with the YOLO architecture described earlier:

- Residual blocks: A number of grids are originally placed throughout the image. The size of each grid is S×S. The graphic in Figure 9.9 shows the grids that were produced using an input image. In Figure 9.9, there are many grid cells with identical size. Each grid cell will have the ability to recognize objects that enter it. For instance, if an item's center falls within a certain grid cell, that cell will be responsible for identifying the object.
- Bounding box regression: A bounding box provides as an outline in a photograph to highlight a specific object. Height (bh), Width (bw), and Class are the attributes of each bounding box in the image (e.g., person, automobile, traffic light). These characteristics are represented by the letter C and the center of the bounding box (bx, by). YOLO calculates an item's height, width, center, and class using a single bounding box regression. Figure 9.9 depicts the likelihood of an object fitting within the bounding box. Box overlapping can be explained using the object detection phenomenon known as intersection over union (IOU). YOLO employs IOU to construct an output box that properly encloses the objects.

When compared to all pre-trained models, the YOLO architecture is noted for operating quite quickly. Due to its algorithm and training properties, it has a high generative and generalized network. It has the unique capacity to process each frame at a rate ranging from 150 frames per second (fps), which is a smaller network and better suited for using the model for real-time applications, to 45 frames per second (fps), which is a faster network. Detecting small items from a picture is quite challenging, and, in some instances, it will not work because the detected objects are too close to one another on the grid. It is recommended to upgrade to the most recent versions of YOLO in order to gain better outcomes when improvising and achieving high accuracy with performance using the limitations of the YOLO architecture.

9.3   SUMMARY

This chapter includes a detailed analysis of pre-trained models and covers every aspect of CV in deep learning. The pre-trained mode is used in CV to simplify the work so that it may be more readily integrated with the application, to quickly obtain a strong model performance, and to process data without the requirement for well-maintained labeled data. The chapter also covers the use of pre-trained models in well-known advancements. The pre-trained architectures are covered including LeNet, AlexNet, R-CNN, Fast R-CNN, Faster R-CNN, Inception, Mask R-CNN, and YOLO. Each pre-trained concept's architecture was described, along with a specific architectural example showing the working. The following chapter will concentrate on current computer-vision-based applications.

BIBLIOGRAPHY

Girshick, R., Donahue, J., Darrell, T., & Malik, J. (2014). Rich feature hierarchies for accurate object detection and semantic segmentation. In *Proceedings of the IEEE conference on computer vision and pattern recognition* (pp. 580–587). New York: IEEE.

Girshick, R. (2015). Fast R-CNN. In *Proceedings of the IEEE international conference on computer vision* (pp. 1440–1448). New York: IEEE

He, K., Gkioxari, G., Dollár, P., & Girshick, R. (2017). Mask R-CNN. In *Proceedings of the IEEE international conference on computer vision* (pp. 2961–2969). New York: IEEE.

Krizhevsky, A., Sutskever, I., & Hinton, G. E. (2012). ImageNet classification with deep convolutional neural networks. *Advances in Neural Information Processing Systems*, 25.

LeCun, Y., Bottou, L., Bengio, Y., & Haffner, P. (1998). Gradient-based learning applied to document recognition. *Proceedings of the IEEE*, 86(11), 2278–2324.

Redmon, J., Divvala, S., Girshick, R., & Farhadi, A. (2016). You only look once: Unified, real-time object detection. In *Proceedings of the IEEE conference on computer vision and pattern recognition* (pp. 779–788). New York: IEEE.

Ren, S., He, K., Girshick, R., & Sun, J. (2015). Faster R-CNN: Towards real-time object detection with region proposal networks. *Advances in Neural Information Processing Systems*, 28.

Simonyan, K., & Zisserman, A. (2014). Very deep convolutional networks for large-scale image recognition. *CoRR, abs/1409.1556*.

Szegedy, C., Vanhoucke, V., Ioffe, S., Shlens, J., & Wojna, Z. (2016). Rethinking the inception architecture for computer vision. In *Proceedings of the IEEE conference on computer vision and pattern recognition* (pp. 2818–2826). New York: IEEE.

Voulodimos, A., Doulamis, N., Doulamis, A., & Protopapadakis, E. (2018). Deep learning for computer vision: A brief review. *Computational Intelligence and Neuroscience, 2018*.

10 Applications of Computer Vision

LEARNING OUTCOMES

After reading this chapter, you will be able to:

- Understand basic knowledge, theories, and techniques in image processing and CV.
- Identify and solve problems in CV and analyze and examine existing practical CV systems.
- Get an exposure of practical application of CV.

10.1 INTRODUCTION

Digital images, videos, and other visible inputs may all be used as sources of relevant information by machines and systems owing to CV. AI aids in the way computers think, whereas CV aids in how they observe and comprehend their surroundings. CV is being used in more areas than was expected. CV has become a part of our daily lives, from identifying early cancer signals to enabling automated checkouts in stores. Here are a few additional uses for CV:

- Optical character recognition
- Image captioning
- Face detection
- Posture detection

10.2 OPTICAL CHARACTER RECOGNITION

Optical character recognition (OCR) is a method for extracting text from images and turning it into machine-readable text which Python programmers can access and alter as a text expression (or any other computer language). However, the picture can be changed into a text record with its items saved as text information utilizing OCR.

Using OCR technology, which separates letters from images, transforms those like letters, and then puts the words in sentences, the original content may be retrieved and changed. Furthermore, it does away with the requirement for human data entry. OCR systems transform printed documents into a text that can be read by a machine by combining hardware and software.

Hardware, such as an imaging device or specialized connecting wires, is used to copy or read text; the additional processing is again often performed by software. In order to construct more complex intelligent character recognition (ICR) approaches,

DOI: 10.1201/9781003348689-10

such as identifying languages or handwritten styles, AI techniques can be incorporated. The main reason OCR is used is so that users may format, modify, and search texts.

Some of the widely accepted frameworks are as follows:

Language used: Python
Framework used: Keras using TensorFlow backend and implemented on Jupyter notebook, Anaconda IDE.
Dataset used
MNIST ("Modified National Institute of Standards and Technology")

Since its 1999 publication, this well-known library of handwritten pictures has served as the standard for classification algorithms as shown in Figure 10.1. MNIST continues to be a trustworthy resource alike even as new machine learning methodologies are developed.

10.2.1 CODE SNIPPETS

```
import tensorflow as tf
(x_train, y_train), (x_test, y_test) = tf.keras.datasets.mnist.load_data()
```

Here, the training and evaluation datasets are extracted from the data. The dataset consists of 60,000 photos for trained and 10,000 testing images. Grayscale codes are used in the x train and x test, while labels representing the numbers 0 through 9 are used in the y test and y train. When examining the form of datasets to determine whether they can be used with CNN or not, the result shown is (60000,28,28), which indicates that our dataset has 60,000 photos, each of which is 28×28 pixels in size. A four-dimensional array is required to use the Keras API, but only a 3D NumPy array is used.

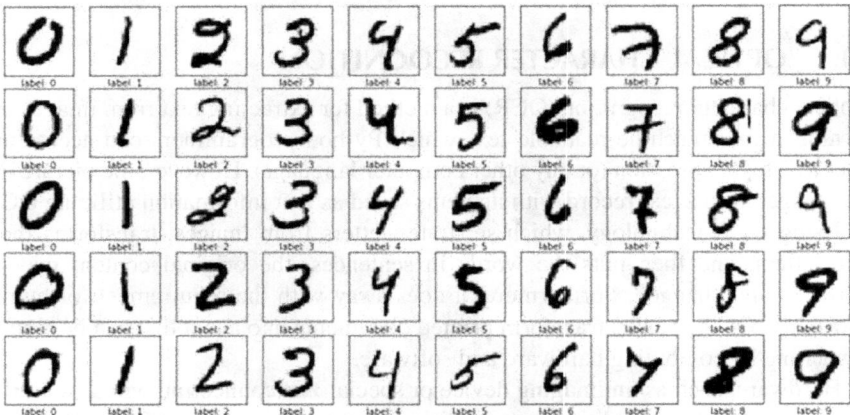

FIGURE 10.1   Labels in MNIST Dataset.

```
x_train = x_train.reshape(x_train.shape[0], 28, 28, 1)
x_test = x_test.reshape(x_test.shape[0], 28, 28, 1)
input_shape = (28, 28, 1)
```

In order to have floating-point values after the division, we set the type of the four-dimensional NumPy array to float.

```
x_train = x_train.astype('float32')
x_test = x_test.astype('float32')
```

Now that we've reached the normalization stage, we always do it in our neural network models. To do this, we divide it by 255.

```
x_train = x_train / 255
x_test = x_test / 255
```

```
from tensorflow.keras.models import Sequential
from tensorflow.keras.layers import Dense,Conv2D,Dropout,Flatten,
MaxPooling2D
model_name=Sequential()
model_name.add(Conv2D(28,kernel size=(3,3),input shape=input shape))
model_name.add(MaxPooling2D(pool_size=(2,2)))
model_name.add(Flatten())
model_name.add(Dense(128,activation=tf.nn.relu))
model_name.add(Dropout(0.2))
model_name.add(Dense(10,activation=tf.nn.softmax))
```

Built the model using the Keras API. Conv2D, MaxPooling, Flatten, Dropout, and Dense layers to be added after importing the sequential model from Keras.

While training, dropout layers combat overfitting by ignoring some of the neurons. Before constructing the completely connected layers, flattened layers convert 2D arrays to 1D arrays.

```
model.compile(optimizer='adam', loss='sparse_categorical_crossentropy', metrics=
['accuracy'])
model.fit(x=x_train,y=y_train, epochs=10)
    model.evaluate(x_test, y_test)
```

10.2.2 RESULT ANALYSIS

```
model.evaluate(x_test, y_test)

313/313 [==============================] - 2s 7ms/step - loss: 0.0584 - accuracy: 0.9859
[0.05842180177569389, 0.9858999848365784]
```

```
model.evaluate(x_train, y_train)

1875/1875 [==============================] - 15s 8ms/step - loss: 0.0238 - accuracy: 0.9925
[0.023804431781172752, 0.9925333261489868]
```

Check the prediction

```
image_index = 2853
plt.imshow(x_test[image_index].reshape(28, 28),cmap='Greys')
predict = x_test[image_index].reshape(28,28)
pred = model.predict(x_test[image_index].reshape(1, 28, 28, 1))
print(pred.argmax())
```

Output
7

Here, we have accessed an image of a character which is "number 7," and the model has predicted accurately.

10.3   FACE AND FACIAL EXPRESSION RECOGNITION

10.3.1   FACE RECOGNITION

In order to verify or uniquely verify identity, software that recognizes faces analyzes and examines patterns based on that individual's face features. Security is where face recognition is most commonly used, although applications in other fields as well are expanding. The fact that facial recognition technology has a wide variety of uses in both police forces and other industries has really attracted a lot of interest.

10.3.2   FACIAL RECOGNITION SYSTEM

There are several techniques used for facial identification, such as the generalized matching face detection strategy and the adaptive local blend linear correlation. Numerous nodal observations on the human face serve as the foundation for the majority of facial recognition systems. A person's identity can be verified or confirmed using the variables associated with their facial point. Using this technique, applications can quickly and accurately recognize the intended users using collected facial data. Image recognition systems are constantly developing and fixing issues with earlier approaches, thanks to new techniques like 3D modelling.

There are several advantages to using face recognition. Especially compared to other options, recognition software is a noncontact fingerprint identification technique. Face photographs may be taken and assessed without the client or subject ever needing to be talked to. As a result, no one is able to mimic another person. Facial recognition is an excellent security solution for timekeeping and attendance monitoring. Since facial recognition technology requires less processing than other biometric techniques, it is also less costly.

10.3.3   MAJOR CHALLENGES IN RECOGNIZING FACE EXPRESSION

10.3.3.1   Illumination

Illumination is the term used to describe variations in light. Automatic facial recognition is a big challenge since even a slight change in illumination can have a large impact on the findings. If the lighting is unexpected, the same subject is captured utilizing the same detector, and they are in virtually identical positions and facial expressions, the results may appear significantly different.

10.3.3.2   Pose Variation

Stance facial recognition algorithms are quite reactive to pose changes. The attitude of a person's face varies as their shaft rotates or their view of the situation changes. Head motions or various camera angles can constantly modify how a face appears, causing intra-class variations and a dramatic drop in automatic face recognition results.

10.3.3.3   Occlusion

An obstruction known as occlusion occurs when one or perhaps more facial characteristics are hidden and the full face is not accessible for input. Occlusion is one of the main issues with facial recognition technology. It frequently occurs in real-world circumstances and is triggered by accessories, beards, and moustaches (such as goggles, hats, and masks). These components diversify the subject, which makes computerized biometric technology a difficult issue to resolve.

10.3.3.4   Expression

The face represents one of the most essential biometrics since it is so crucial to a personality and feelings. Different situations lead to various moods, which in turn lead people to display various emotions and, eventually, alter their facial expressions. The many ways which the same individual thinks themselves are a crucial factor as well. Human emotions including joy, sorrow, fury, contempt, fear, and surprise are all examples of macro-expressions. Facial cues are brief, unplanned alterations in the direction the face moves.

10.3.3.5   Low Resolution

A common image should have at least a 16×16 resolution. Images with less than 16×16 resolution are considered of low quality. These low-resolution images may be found to be used in centric view cameras, such as CCTV cameras and grocery store security cameras. Due to the distance between the cameras and the face, they can only capture a limited (16×16) section of the human face. There are not many details in a photograph with very low resolution.

10.3.3.6   Ageing

The fact that the texture and look of the face vary over time and that ageing is reflected in them present another challenge for facial recognition systems. As we age, our faces' features, contours, and lines alter, among other things. For accuracy testing, a dataset is created for a range of age groups throughout time. It is done for long-term picture retrieval and visual inspection. The recognition procedure in this instance depends heavily on image retrieval, which makes use of essential characteristics like creases, marks, eyebrows, and hairstyles.

10.3.3.7   Model Complexity

The CNN architecture used by current facial recognition technologies is "very deep", "overly complicated," and unsuited for including the practice on embedded devices.

Changes in illumination, emotion, posture, and occlusion must be accommodated by a face recognition system. With a need to take a minimal number of photos during registration and the elimination of sophisticated architecture, it should be adaptive to a large number of users.

10.3.3.8 Framework Used

Language used: Python

Framework used: Keras using TensorFlow backend and implemented on Jupyter notebook, Anaconda IDE.

10.3.3.9 Dataset

A library of face images called the Labelled Faces in the Wild (LFW) dataset was created to examine the issue of unrestricted face identification. The size of the dataset is 173 MB and contains over 13,000 face images that were gathered from the Internet.

10.3.3.10 Code Snippets

The image is captured through a camera connected to the machine, and the captured image is converted from BGR to RGB color. Then the model processes each frame of the video to plot the landmarks on the face.

```
with mp_holistic.Holistic(min_detection_confidence=0.5,min_tracking_confi-
dence=0.5) as hol
    while(True):
    ret, frame = cap.read()
    #Recoloring the feed
    image = cv2.cvtColor(frame, cv2.COLOR_BGR2RGB)
    results = holistic.process(image)
    print(results)
```

10.3.4 Result Analysis

The face is detected and recognized. The facial key correspond to a particular face feature, such as nose tip, and the center of eyes as shown in figure 10.2.

10.4 VISUAL-BASED GESTURE RECOGNITION

A gesture is a deliberate physical movement made with the hands, arms, fingers, or other body parts in order to express meaning or information to the environment. A good interpretation of the human hand as meaningful commands is necessary for gesture recognition as shown in figure 10.3. There are several ways to categorize vision-based methods for displaying, segmenting, and identifying human behaviors.

10.4.1 Framework Used

Language used: Python.

Frameworks used: MediaPipe, backend and implemented on Jupyter notebook, and Anaconda IDE.

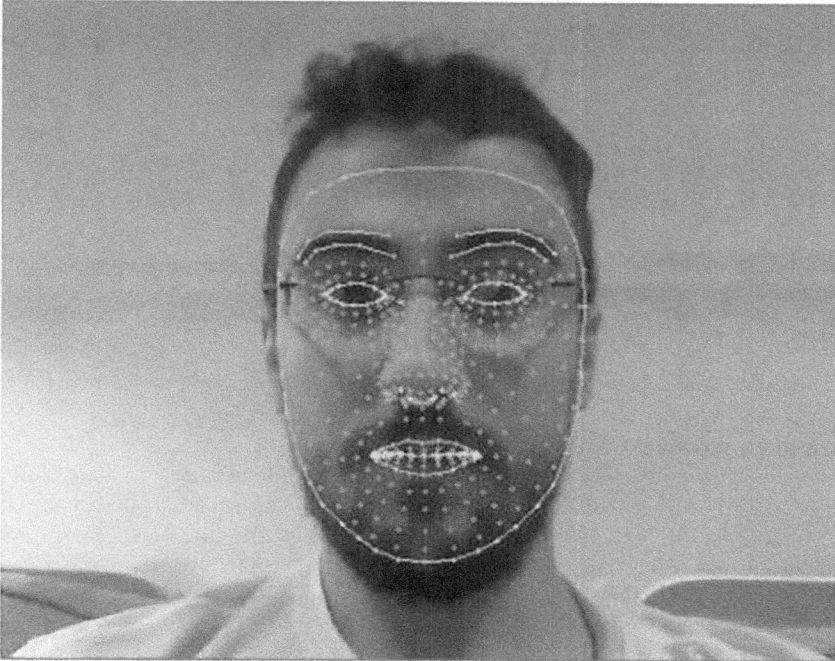

FIGURE 10.2   Face Recognition.

Dataset: Jester dataset

Using a webcam, 148,092 annotated video clips of people making simple, pre-defined hand gestures are included in the Jester gesture detection dataset. It is intended to educate computer programmers to recognize human hand gestures including swiping left or right, sliding two fingers, and drumming the fingers.

The videos include 27 distinct components of human hand gestures and are divided into training, development, and testing segments in an 8:1:1 ratio. Two "no gesture" classes are included in the dataset to aid the network in differentiating between recognized hand movements and unidentified gestures.

Gesture recognition as well as its function in human–computer interactions has become more significant in the era of mobile computing. Using the Jester video collection, reliable machine learning models can be trained to recognize human hand movements.

10.4.2   CODE SNIPPETS

First, in the following code, we will first initialize a holistic model and draw the utilities using MediaPipe.

```
mp_Holistic = mp.Solution.Holistic # Holistic model
mp_draw = mp.Solution.draw_utils # Draw utilities
```

The mediapipe_dect function is used to draw the mediapipe model on the image. Initially, the image is converted from BGR to RGB image, and the flags are made non-writable. After processing the image, the flags are again made writable. The image is then converted from RGB to BGR as a result.

```
def mediapipe_dect(image, model):
image = cv2.cvtColor(image, cv2.COLOR_BGR2RGB) image.flags.writeable =
    False
model.process(image) image.flags.writeabe = True
image = cv2.cvtColor(image, cv2.COLOR_RGB2BGR)
return image, results
```

The draw_landmark function is for plotting the landmarks on the given image.

```
def draw_landmark(image, results):
mp_draw.draw_landmark(image,        results.face_landmark,        mp_Holistic.
    FACEMESH_TESSELAT mp_draw.draw_landmark(image, results.pose_land-
    mark, mp_Holistic.POSE_CONNECTIONS) mp_draw.draw_landmark(image,
    results.left_hand_landmark, mp_Holistic.HAND_CONNECT mp_draw.draw_
    landmark(image, results.right_hand_landmark, mp_Holistic.HAND_CONNEC
```

The following function is to provide styles for the drawn landmarks. For example, the landmarks drawn on hands and legs are of different colors to differentiate from each other.

```
def draw_styled_landmark(image, results):
    # face connections
    mp_draw.draw_landmark(image, results.face_landmark, mp_Holistic.FACEMESH_
    TESSELAT mp_draw.DrawSpec(color=(80,110,10), thickness=1, circle
    mp_draw.DrawSpec(color=(80,256,121), thickness=1, circle) # pose connections
mp_draw.draw_landmark(image,        results.pose_landmark,        mp_Holistic.POSE_
CONNECTIONS, mp_draw.DrawSpec(color=(80,22,10), thickness=2, circle_ mp_
draw.DrawSpec(color=(80,44,121), thickness=2, circle)
    # left hand connections
mp_draw.draw_landmark(image, results.left_hand_landmark, mp_Holistic.HAND_
CONNECT mp_draw.DrawSpec(color=(121,22,76), thickness=2, circle mp_draw.
DrawSpec(color=(121,44,250), thickness=2, circle)
    # right hand connections
mp_draw.draw_landmark(image, results.right_hand_landmark, mp_Holistic.HAND_
CONNEC mp_draw.DrawSpec(color=(245,117,66), thickness=2,

cap = cv2.VideoCapture(0)
# Set mediapipe model
with mp_holistic.Holistic(min_detection_confidence=0.5, min_tracking_confi-
    dence=0.5) as holistic:
while cap.isOpened():
# Read feed
```

```
ret, frame = cap.read()
# Make detections
image, results = mediapipe_detection(frame, holistic)
print(results)
# Draw landmarks
draw_styled_landmarks(image, results)
# Show to screen
cv2.imshow('OpenCV Feed', image)
# Break gracefully
if cv2.waitKey(10) & 0xFF == ord('q'):
break
cap.release()
cv2.destroyAllWindows()
draw_landmarks(frame, results)
plt.imshow(cv2.cvtColor(frame, cv2.COLOR_BGR2RGB))
len(results.right_hand_landmarks.landmark)
pose = []
for res in results.pose_landmarks.landmark:
test = np.array([res.x, res.y, res.z, res.visibility])
pose.append(test)
pose= np.array([[res.x, res.y, res.z, res.visibility] for res in results.pose_landmarks.
    landmark]).flatten()
pose = np.array([[res.x, res.y, res.z, res.visibility] for res in results.pose_land-
    marks.landmark]).flatten() if results.pose_landmarks else np.zeros(132)
face = np.array([[res.x, res.y, res.z] for res in results.face_landmarks.landmark]).
    flatten() if results.face_landmarks else np.zeros(1404)
lh = np.array([[res.x, res.y, res.z] for res in results.left_hand_landmarks.land-
    mark]).flatten() if results.left_hand_landmarks else np.zeros(21*3)
rh = np.array([[res.x, res.y, res.z] for res in results.right_hand_landmarks.land-
    mark]).flatten() if results.right_hand_landmarks else np.zeros(21*3)
rh
def extract_keypoints(results):
pose = np.array([[res.x, res.y, res.z, res.visibility] for res in results.pose_land-
    marks.landmark]).flatten() if results.pose_landmarks else np.zeros(33*4)
face = np.array([[res.x, res.y, res.z] for res in results.face_landmarks.landmark]).
    flatten() if results.face_landmarks else np.zeros(468*3)
lh = np.array([[res.x, res.y, res.z] for res in results.left_hand_landmarks.land-
    mark]).flatten() if results.left_hand_landmarks else np.zeros(21*3)
rh = np.array([[res.x, res.y, res.z] for res in results.right_hand_landmarks.land-
    mark]).flatten() if results.right_hand_landmarks else np.zeros(21*3)
return np.concatenate([pose, face, lh, rh])
result_test = extract_keypoints(results)
result_test
np.save('0', result_test)
np.load('0.npy')
```

```
# Path for exported data, numpy arrays
DATA_PATH = os.path.join('MP_Data')
# Actions that we try to detect
actions = np.array(['hello', 'thanks', 'drink'])
# Thirty videos worth of data
no_sequences = 30
# Videos are going to be 30 frames in length
sequence_length = 30
# Folder start
start_folder = 30
for action in actions:
for sequence in range(no_sequences):
try:
os.makedirs(os.path.join(DATA_PATH, action, str(sequence)))
except:
pass
cap = cv2.VideoCapture(0)
# Set mediapipe model
with mp_holistic.Holistic(min_detection_confidence=0.5, min_tracking_confi-
    dence=0.5) as holistic:
# Loop through actions
for action in actions:
# Loop through sequences aka videos
for sequence in range(no_sequences):
# Loop through video length aka sequence length
for frame_num in range(sequence_length):
# Read feed
ret, frame = cap.read()
# Make detections
image, results = mediapipe_detection(frame, holistic)
print(results)
# Draw landmarks
draw_styled_landmarks(image, results)
# NEW Apply wait logic
if frame_num == 0:
cv2.putText(image, 'STARTING COLLECTION', (120,200),
cv2.FONT_HERSHEY_SIMPLEX, 1, (0,255, 0), 4, cv2.LINE_AA)
cv2.putText(image, 'Collecting frames for {} Video Number {}'.format(action,
    sequence), (15,12),
cv2.FONT_HERSHEY_SIMPLEX, 0.5, (0, 0, 255), 1, cv2.LINE_AA)
#show to screen
cv2.imshow('OpenCV Feed',image)
cv2.waitKey(1000)
else:
cv2.putText(image, 'Collecting frames for {} Video Number {}'.format(action,
    sequence), (15,12),
```

```
cv2.FONT_HERSHEY_SIMPLEX, 0.5, (0, 0, 255), 1, cv2.LINE_AA)
#show to screen
cv2.imshow('OpenCV Feed',image)
# NEW Export keypoints
keypoints = extract_keypoints(results)
npy_path = os.path.join(DATA_PATH, action, str(sequence), str(frame_num))
np.save(npy_path, keypoints)
# Break gracefully
if cv2.waitKey(10) & 0xFF == ord('q'):
break
cap.release()
cv2.destroyAllWindows()
from sklearn.model_selection import train_test_split
from tensorflow.keras.utils import to_categorical
label_map = {label:num for num, label in enumerate(actions)}
label_map
sequences, labels = [], []
for action in actions:
for sequence in range(no_sequences):
window = []
for frame_num in range(sequence_length):
res = np.load(os.path.join(DATA_PATH, action, str(sequence), "{}.npy".
    format(frame_num)))
window.append(res)
sequences.append(window)
labels.append(label_map[action])
np.array(sequences).shape
np.array(labels).shape
X = np.array(sequences)
X.shape
y = to_categorical(labels).astype(int)
y
X_train, X_test, y_train, y_test = train_test_split(X, y, test_size=0.05)
X_train.shape
from tensorflow.keras.models import Sequential
from tensorflow.keras.layers import LSTM, Dense
from tensorflow.keras.callbacks import TensorBoard
log_dir = os.path.join('Logs')
tb_callback = TensorBoard(log_dir=log_dir)
model = Sequential()
model.add(LSTM(64, return_sequences=True, activation='relu',
    input_shape=(30,1662)))
model.add(LSTM(128, return_sequences=True, activation='relu'))
model.add(LSTM(64, return_sequences=False, activation='relu'))
model.add(Dense(64, activation='relu'))
model.add(Dense(32, activation='relu'))
```

```
model.add(Dense(actions.shape[0], activation='softmax'))
model.compile(optimizer='Adam',                loss='categorical_crossentropy',
    metrics=['categorical_accuracy'])
model.fit(X_train, y_train, epochs=100, callbacks=[tb_callback])
model.summary()
res = model.predict(X_test)
actions[np.argmax(res[1])]
actions[np.argmax(y_test[1])]
model.save('action.h5')
from sklearn.metrics import multilabel_confusion_matrix, accuracy_score
yhat = model.predict(X_test)
ytrue = np.argmax(y_test, axis=1).tolist()
yhat = np.argmax(yhat, axis=1).tolist()
multilabel_confusion_matrix(ytrue, yhat)
accuracy_score(ytrue, yhat)
from scipy import stats
colors = [(245,117,16), (117,245,16), (16,117,245)]
def prob_viz(res, actions, input_frame, colors):
output_frame = input_frame.copy()
for num, prob in enumerate(res):
cv2.rectangle(output_frame, (0,60+num*40), (int(prob*100), 90+num*40), col-
    ors[num], -1)
cv2.putText(output_frame,    actions[num],    (0,    85+num*40),    cv2.FONT_
    HERSHEY_SIMPLEX, 1, (255,255,255), 2, cv2.LINE_AA)
return output_frame
plt.figure(figsize=(18,18))
#plt.imshow(prob_viz(res, actions, image, colors))
# 1. New detection variables
sequence = []
sentence = []
threshold = 0.5
cap = cv2.VideoCapture(0)
# Set mediapipe model
with  mp_holistic.Holistic(min_detection_confidence=0.5,  min_tracking_confi-
    dence=0.5) as holistic:
while cap.isOpened():
# Read feed
ret, frame = cap.read()
# Make detections
image, results = mediapipe_detection(frame, holistic)
print(results
# Draw landmarks
draw_styled_landmarks(image, results)
# 2. Prediction logic
```

```
keypoints = extract_keypoints(results)
sequence.insert(0,keypoints)
sequence = sequence[:30]
if len(sequence) == 30:
res = model.predict(np.expand_dims(sequence, axis=0))[0]
print(actions[np.argmax(res)])
#3. Viz logic
if res[np.argmax(res)] > threshold:
if len(sentence) > 0:
if actions[np.argmax(res)]!= sentence[-1]:
sentence.append(actions[np.argmax(res)])
else:
sentence.append(actions[np.argmax(res)])
if len(sentence) > 5:
sentence = sentence[-5:]
# Viz probabilities
image = prob_viz(res, actions, image, colors)
cv2.rectangle(image, (0,0), (640, 40), (245, 117, 16), -1)
cv2.putText(image, ''.join(sentence), (3,30),
cv2.FONT_HERSHEY_SIMPLEX, 1, (255, 255, 255), 2, cv2.LINE_AA)
# Show to screen
cv2.imshow('OpenCV Feed', image)
# Break gracefully
if cv2.waitKey(10) & 0xFF == ord('q'):
break
cap.release()
cv2.destroyAllWindows()
```

10.4.3   RESULT ANALYSIS

The hand gesture is recognized, and the correct output for the gesture is obtained.

10.4.4   MAJOR CHALLENGES IN GESTURE RECOGNITION

10.4.4.1   Uncommon Backgrounds

Any location should allow gesture recognition to work well; it should work if you're already walking, in a car, or at house. Machine learning may be used to teach a machine to reliably discriminate between the finger and the backdrop.

10.4.4.2   Movement

Common sense dictates that a gesture corresponds more to a movement than a static picture. Therefore, pattern detection for gestures should be possible. To shut the current application, for example, we may recognize a wave pattern as a command rather than just an image of a hand extended.

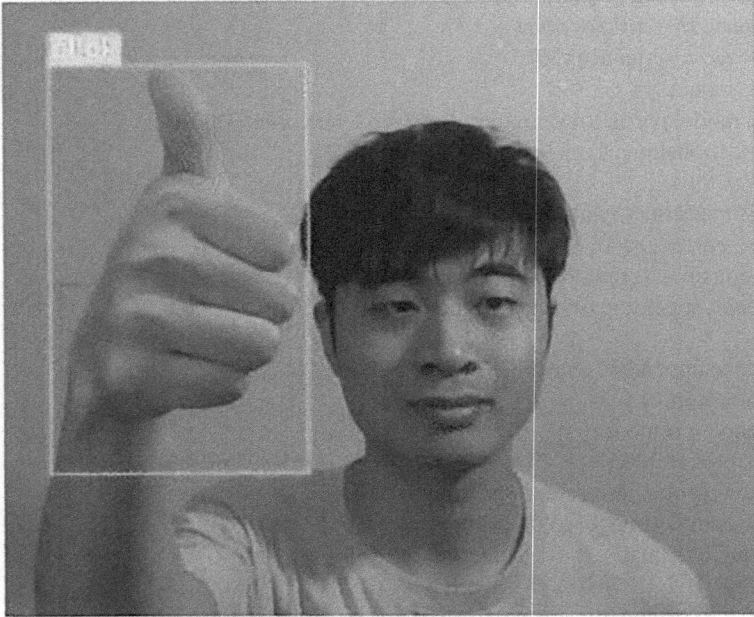

FIGURE 10.3   Gesture Recognition.

10.4.4.3 Combination of Motions

Furthermore, gestures might consist of a variety of actions, therefore it's important to offer the context in order to recognize patterns like rotational finger gestures and thumb presentations that can be used to denote a certain space or a limited number of files.

10.4.4.4 A Range of Gestures

There are many different ways that humans may make different actions. Although we are relatively tolerant of errors, this unpredictable nature may make it increasingly challenging for robots to classify gestures.

10.4.4.5 Overcoming the Latency

The motion detection method must eliminate the lag between making motions and its categorization. Showing how rapid and accurate hand gestures can be is the only way to encourage the usage of them. There isn't really any other reason to start using gestures if they don't quicken and simplify your engagement. We want as little latency as possible so that we can truly provide the customer with feedback right away.

10.5 POSTURE DETECTION AND CORRECTION

Interpreting sign language, managing full-body movements, and tracking physical activity all depend on being able to predict the human location from videos. Postures are the basis for exercises in yoga, dancing, and fitness, for example. Techniques like posture detection can also enable the projection of digital data and content over the physical environment in virtual worlds.

A significantly high- position model called Blaze Pose was created primarily to support difficult areas like yoga, fitness, and dance. It extends the 17 key-point topology of the initial Pose Net model that we launched a few years ago by being able to identify 33 key points. These extra key points offer crucial details regarding the position of the face, hands, and feet along with scale and rotation.

10.5.1 Framework Used

Language used: Python

Framework used: MediaPipe, Blaze poses, backend and implemented on Jupiter notebook, Anaconda IDE.

MediaPipe and other necessary packages needed should be installed the dimension of the image should be provided. And read the image once to preview it.

10.5.2 Squats

The following code snippet can be used to detect postures involved in squats exercise:

```
!pip install mediapipe
import cv2
from google.colab.patches import cv2_imshow
import math
import numpy as np
DESIRED_HEIGHT = 480
DESIRED_WIDTH = 480
def resize_and_show(image):
h, w = image.shape[:2]
if h < w:
img=cv2.resize(image,(DESIRED_WIDTH,math.floor(h/(w/DESIRED_WIDTH))))
else:
img = cv2.resize(image, (math.floor(w/(h/DESIRED_HEIGHT)), DESIRED_
    HEIGHT))
cv2_imshow(img)
# Read images with OpenCV.
images = {name: cv2.imread(name) for name in uploaded.keys()}
# Preview the images.
for name, image in images.items():
print(name)
resize_and_show(image)
import mediapipe as mp
mp_pose = mp.solutions.pose
mp_drawing = mp.solutions.drawing_utils
mp_drawing_styles = mp.solutions.drawing_styles
help(mp_pose.Pose)
```

```python
with mp_pose.Pose(
static_image_mode=True,     min_detection_confidence=0.5,     model_complex-
    ity=2) as pose:
for name, image in images.items():
# Convert the BGR image to RGB and process it with MediaPipe Pose.
results = pose.process(cv2.cvtColor(image, cv2.COLOR_BGR2RGB))
# Print nose landmark.
image_hight, image_width, _ = image.shape
if not results.pose_landmarks:
continue
print(
f'Nose coordinates: ('
f'{results.pose_landmarks.landmark[mp_pose.PoseLandmark.
    NOSE].x * image_width}, '
f'{results.pose_landmarks.landmark[mp_pose.PoseLandmark.
    NOSE].y * image_hight})'
)
# Draw pose landmarks.
print(f'Pose landmarks of {name}:')
annotated_image = image.copy()
mp_drawing.draw_landmarks(
annotated_image,
results.pose_landmarks,
mp_pose.POSE_CONNECTIONS,
landmark_drawing_spec=mp_drawing_styles.get_default_pose_landmarks_
    style())
resize_and_show(annotated_image)
import math
def getAngle(firstPoint, midPoint, lastPoint):
result      =math.degrees(math.atan2(lastPoint.y—midPoint.y,lastPoint.x—mid-
    Point.x)-math.atan2(firstPoint.y—midPoint.y,firstPoint.x—midPoint.x))
result = abs(result)# Angle should never be negative
if (result > 180):
result = (360.0—result)# Always get the acute representation of the angle
return result
def distance_between(a,b):
a = np.array(a)
b= np.array(b)
return np.linalg.norm(a—b)
with mp_pose.Pose(
static_image_mode=True,     min_detection_confidence=0.5,     model_complex-
    ity=2) as pose:
for name, image in images.items():
results = pose.process(cv2.cvtColor(image, cv2.COLOR_BGR2RGB))
# Print the real-world 3D coordinates of nose in meters with the origin at
# the center between hips.
```

```
print("Angle between left hip,Knee,ankle: {}".format(getAngle(results.pose_
    world_landmarks.landmark[mp_pose.PoseLandmark.LEFT_HIP],results.
    pose_world_landmarks.landmark[mp_pose.PoseLandmark.LEFT_
    KNEE],results.pose_world_landmarks.landmark[mp_pose.PoseLandmark.
    LEFT_ANKLE])))
angle          =          getAngle(results.pose_world_landmarks.landmark[mp_pose.
    PoseLandmark.LEFT_HIP],results.pose_world_landmarks.landmark[mp_
    pose.PoseLandmark.LEFT_KNEE],results.pose_world_landmarks.land-
    mark[mp_pose.PoseLandmark.LEFT_ANKLE])
if(angle<80–20):
print("Sqaut less")
elif(angle > 80+20):
print("Sqaut more")
else:
print("Perfect")
angle=getAngle(results.pose_world_landmarks.landmark[mp_pose.
    PoseLandmark.LEFT_SHOULDER],results.pose_world_landmarks.land-
    mark[mp_pose.PoseLandmark.LEFT_HIP],results.pose_world_landmarks.
    landmark[mp_pose.PoseLandmark.LEFT_KNEE])
#print(angle)
if(angle < 90):
print("Straighten your back")
elif(angle==90):
print("Perfect")
else:
print("Squat more")
ang = getAngle(results.pose_world_landmarks.landmark[mp_pose.PoseLa
if(ang<=90):
print("Perfect")
else:
print("Straighten your toe")
if(ang<=90):
print("Perfect")
else:
print("Straighten your toe")
if(ang<=90):
print("Perfect")
else:
print("Straighten your toe") width . . .")
hand          =          getAngle(results.pose_world_landmarks.landmark[mp_pose.
    PoseLandmark.LEFT_WRIST],results.pose_world_landmarks.landmark[mp_
    pose.PoseLandmark.LEFT_ELBOW],results.pose_world_landmarks.land-
    mark[mp_pose.PoseLandmark.LEFT_SHOULDER])
#print(results.pose_world_landmarks.landmark[mp_pose.PoseLandmark.
    LEFT_ANKLE].y)
```

The results of the pose detection code is depicted in Figure 10.4.

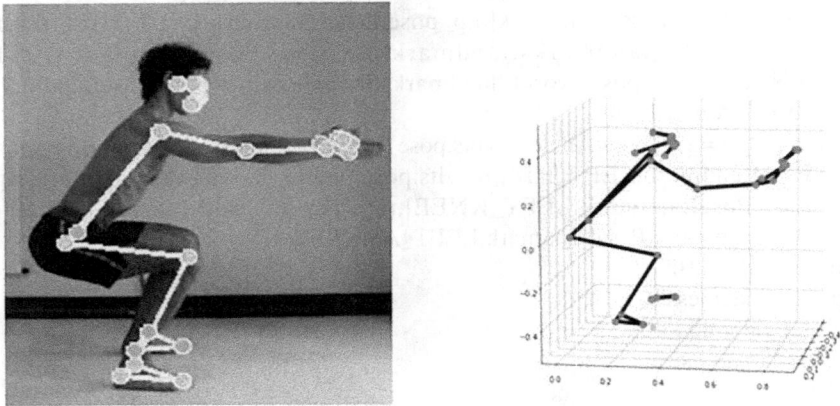

FIGURE 10.4   (a) Pose Detection and (b) Pose Detection Plot.

10.5.3 RESULT ANALYSIS

In the end, we built a classifier to classify yoga poses using MediaPipe. The pose is correctly detected, and the key points are clearly visible. This can be processed on other yoga poses.

10.6 SUMMARY

This chapter explains some of the real-world applications of CV. We have used many libraries and datasets for different applications. This helps the reader to identify the problems and solve using CV and build models using appropriate open-source frameworks. CV is being used in more and more applications every day. Numerous datasets are accessible, and these can train computers to recognize and understand items. This technology also exemplifies a crucial development in our civilization's quest to develop AI which will match human intelligence.

BIBLIOGRAPHY

El-Komy, A., Shahin, O. R., Abd El-Aziz, R. M., & Taloba, A. I. (2022). Integration of computer vision and natural language processing in multimedia robotics application. *Information Sciences, 7*, 6.

Li, Y., & Zhang, Y. (2020). Application research of computer vision technology in automation. *2020 International Conference on Computer Information and Big Data Applications (CIBDA)* (pp. 374–377). New York: IEEE. doi: 10.1109/CIBDA50819.2020.00090.

Long, T., Gao, Q., Xu, L., & Zhou, Z. (2022). A survey on adversarial attacks in computer vision: Taxonomy, visualization and future directions. *Computers & Security*, 102847.

Messaoud, S., Bouaafia, S., Maraoui, A., Ammari, A. C., Khriji, L., & Machhout, M. (2022). Deep convolutional neural networks-based hardware–Software on-chip system for computer vision application. *Computers & Electrical Engineering, 98*, 107671.

Sinha, G. R., Subudhi, B., Fan, C. P., & Das, D. (2022). Attaining strong learning outcomes using modern pedagogies in teaching image processing and computer vision. In *Development of Employability Skills through Pragmatic Assessment of Student Learning Outcomes* (pp. 1–20). London: IGI Global.

Xiaogang, W., Siwen, Q., Ji, Z., Junjun, G., & Cao, P. (2022). Exploration and application of library automatic book inventory checking system based on computer vision and artificial intelligence. *Library Journal, 41*(7), 96.

Index

For Product Safety Concerns and Information please contact our EU
representative GPSR@taylorandfrancis.com
Taylor & Francis Verlag GmbH, Kaufingerstraße 24, 80331 München, Germany